Gorillas among Us

Gorillas among Us

A Primate Ethnographer's Book of Days

Dawn Prince-Hughes

Illustrations by Anne Hulse

The University of Arizona Press   Tucson

The University of Arizona Press
First Printing

⊛ This book is printed on acid-free, archival-quality paper.
Manufactured in the United States of America

06   05   04   03   02   01   6   5   4   3   2   1

Library of Congress Cataloging-in-Publication Data
Prince-Hughes, Dawn, 1964–
Gorillas among us : a primate ethnographer's book of days /
Dawn Prince-Hughes ; illustrations by Anne Hulse.
p. cm.
Includes bibliographical references (p.).
ISBN 0-8165-2150-6 (cloth : alk. paper) —
ISBN 0-8165-2151-4 (pbk. : alk. paper)
1. Gorilla—Behavior.   I. Hulse, Anne.   II. Title.
QL737.P96 P74   2001
599.884'15—dc21
2001001464

British Library Cataloguing-in-Publication Data
A catalogue record for this book is available from the
British Library.

For Congo, who was a person.

Contents

Foreword   Jane Goodall

Gorillas among Us is a wonderful book that delights the mind and touches the heart. It tells the story of a researcher who studies a group of gorillas in a zoo. It describes her joy at the first touch of a gorilla, her elation when a baby is born, and her sharing in a sense of communal grief when one of the gorillas dies. The book lays out the life histories of each of the gorillas and explores the interactions and relationships between various individuals. The researcher describes gorilla emotions and shares her belief that the gorillas have an ancient culture.

The author was privileged to experience numerous special events with the gorillas. One day she discovers a pile of stones neatly stacked—almost like a cairn. When she asks the gorilla keeper why the gorillas have done this, the gorilla keeper refuses to speculate on the meaning but says that it has happened before and that she believes it has some special significance for the gorillas.

Prince-Hughes expresses sadness over the fact that zoo visitors, on average, spend so little time watching the gorillas—or any animals in the zoo. Many people stop for only a few brief moments before moving on. Perhaps after reading this book, visitors will be encouraged to spend longer watching the gorillas, not only because these zoo denizens represent one of the animal kingdom's most magnificent species, but also because the visitors will be eager to meet the fascinating personalities that they have read about. And they will understand the meaning of the postures and gestures used by the gorillas in communicating with each other.

It is the author's empathy with her subjects, her obvious love of gorillas, that makes this book so special. The book will serve conservation well, for no one who reads about these wonderful beings, so humanlike yet so uniquely themselves, can fail to be horrified to learn of the plight of wild gorillas in Africa. Their forest habitat is being destroyed as human populations grow. They are caught in snares set for pigs and antelopes, and though they are strong enough to break the wire cable, their struggles pull the noose tight and eventually they lose a hand or foot or die an agonizing death from gangrene. Mothers are shot so that their infants can be stolen for the live animal trade. And in the Congo Basin, the very heart of the gorillas' range, they are hunted—along with chimpanzees, elephants, monkeys, and almost every other animal—for food. Logging companies have built roads deep into the forest, and hunters from towns ride trucks to the ends of the roads, shoot everything, and truck the carcasses back to the cities to sell in the markets, catering to a cultural preference in this part of Africa for the flesh of wild animals. People pay more for gorilla than for goat. This trade, along with the shooting of "bushmeat" for the employees of huge logging camps, is absolutely not sustainable. Unless we can stop the hunting of gorillas for food these magnificent creatures will be almost extinct within the next ten to fifteen years.

Let us pray that this book, portraying gorillas as intelligent beings with a complex society and rich social and emotional lives, with highly developed and rational minds, will make a difference. Prince-Hughes has written a book that speaks to our hearts and enlists our sympathies. Thank you, for the gorillas need us now more than ever before.

Acknowledgments

I want to thank first and most deeply the gorillas I have worked with; their sacrifices great and small have made this book possible.

I also wish to thank Jane Goodall, Violet Sunde, Virginia Landau, James King, William Karesh, Ben Abe, and Kay Eckiss for their professional and, more important, personal support and inspiration.

I thank Woodland Park Zoo, Western Washington University, University of Washington, University of Nairobi, and the Jane Goodall Institute for providing me with research resources.

My family—Davina, Joyce, and Ron Prince; Gregg Alexander; and Bob, Diane, Holly, and Darcy Hughes—have all lent their strength and love to my adventures and have believed in my abilities as a writer.

A special thank you goes to my partner, Tara, who has always stood behind me and who grew to love the gorillas and helped me take the time to write their stories.

Introduction

What follows are stories and reflections based on a decade's worth of diary entries and notes from my research with captive Western lowland gorillas. As early as my first day studying their habits and being near them through their quiet lives, I wanted to tell a part of their story.

As many ethologists have come to know, the stories of the animals we learn from become our own stories, and our stories in turn become the stories of those animals. Some scientists have labeled this reciprocity "anthropomorphism," seeing the process entirely as a projection by humans of what is only human. Such an attitude seems nothing more than a way to turn from a story that never stops unfolding: a story you must sit and listen to forever.

I sat through parts of this story as I joined the gorillas in the cold, the noise, the happy days, and the heartbreaking events that make up a life. Like the names of the ents, the tree-men in J.R.R. Tolkien's *Lord of the Rings* trilogy, gorilla stories are long. At times these stories are too long for human endurance. For a gorilla to tell a story about ancestors takes as long as it took for those ancestors to come into existence; to leave anything out would offend the story.

Of course, human beings leave much of what is important out of their stories, and as a result we have forgotten much of where we came from and what is around us. Because I am only human, my forgetting and lapses helped shape this book. All of the characters we will come to know in these pages (whether gorillas or other people) are composites of many distilled into one. The setting—a zoo much like the one where my research was undertaken—is a composite zoo like all zoos throughout time. However, all the events that unfold in the pages of this book actually happened and were experienced directly by me and other researchers.

Much of the material on wild gorillas contained in this book comes to us from the research of the late Dian Fossey. Like Dian, many of the gorillas I have known are now dead. I have asked them to give me guidance as I tell parts of their stories; I hope I have succeeded in telling them in a more gorillalike way, rather than in the hurried and homogenous way humans are used to.

Many parts of the gorillas' story we share by way of a common ancestor. How we became separate is a part of the story that science has been consumed with; this part of the story is usually told in a human way. Perhaps a more cynical person would suggest that the obsession with telling only this part of the story stems from a deep need to demonstrate our separateness, but I believe there is a returning drive to reconstruct our past, stemming from a primal need to demonstrate our connectedness. This is a story of the threads and the webs, of memories, and of songs.

Gorillas among Us

Prologue

February 27 As I approached the east entrance of the Whitman Pines Zoological Gardens, I looked around anxiously for the gorilla keeper, who had arranged to meet me just inside the walls. It was bitterly cold as the wind swept across a deserted zoo grounds, catching paper cups and cotton candy sticks, rolling them briskly toward me as if to remind me that the August days for visiting the zoo for fun were long over.

I stamped my feet and blew on my hands as the ticket taker watched passively from inside his booth, himself numbed more with boredom than with cold.

"Has anyone been by to ask about me?" I asked the man in the booth.

"I haven't seen anyone since I've been on shift," he said with a yawn.

"How long have you been on?"

"Six hours."

I tried to imagine what six hours would be like in a tiny concrete booth with nothing to do. I guessed that the experience was probably much like the one the animals inside endured all day. I wondered if the ticket taker ever felt like an exhibit. I imagined the obligatory sign above his window: TICKET TAKER: INDIGENOUS TO NORTH AMERICA. ONCE ENDANGERED BUT MAKING A STRONG COMEBACK. PLEASE DO NOT FEED.

I remembered reading about an eccentric millionaire who volunteered once to become an exhibit for a couple of weeks at his local zoo. The proposal did not go over well with the zoo administration. It made sense to me somehow that they would feel comfortable putting someone in a box, but only as long as there was no sign. That would be vulgar.

The ticket taker grunted and pointed a mittened hand in a direction behind me. I turned and saw what I guessed was the gorilla keeper coming slowly toward the booth. She was walking at a brisk pace but stopping often and abruptly to watch the crows flying by and squirrels climbing in and out of garbage cans with absolute absorption.

Because of her punctuated means of locomotion, it was quite some time before she was close enough for me to make out her features. She had black hair with streaks of gray. It blew wildly in the wind, though she did not seem to notice. She was quite short but very solid. Her steps, as she charged along, seemed to root her to the earth each time a foot came down.

When she stopped to watch the birds, she looked for all the world like some rough statue of a rare wild animal. Adding to her unfinished appearance were the torn black coat she wore and the rags of mittens that had bits of raw vegetables stuck to them.

I stood transfixed through the full twenty minutes it took her to make her way over to where I stood. I was seriously wondering if she had forgotten that I was standing there waiting for her in the unbearable cold. She continued to keep her eyes on the crows as she drew up beside me, her shoulder touching mine. She just stood there. As more time went by I began to feel awkward about the invasion of my personal space and reflected that though I had often been called strange, I did not hold a candle to this woman.

Finally, in a long, drawn-out rumble, she cleared her throat. She turned to me with wide brown eyes the color of molasses and, as if she had known me all my life, said, "Dawn . . . it is so good to see you." A gentle smile wrinkled up her face. The creases in her face were unusual somehow, like the lines one might expect to see on museum reconstructions of the faces of old human ancestors. I felt inexplicably related to her as words resonated in my cold ears.

I had the immediate impression that she really had known me all my life and was not only glad to see me, but saw me in a way I was unaccustomed to. It was unnerving but flattering somehow.

She suddenly turned from me and took off walking in a westerly direction. I shot a quick glance back at the ticket taker, who was complacently picking his nose and reading the sports section of the newspaper, then followed her.

"I am Karin Forrmen, the gorilla keeper," she said to me over her shoulder as she sped along with purpose. "You are Dawn, the new graduate student here to study the components of gorilla culture, yes?" It was more of a statement than a question.

"You have an accent. Where are you from?" I puffed out as I struggled along behind her.

"Originally?"

"Yes."

"Africa, of course. As an anthropology student you should know that!" She let out a raspy, panting laugh, barely making any sound. Her eyes watered, and she started coughing. It got worse until she stopped and leaned over the railing in front of the bushpigs, who stopped to watch with intense curiosity her frantic gasping.

"What can I do? Do you need help?" I stared at her, concerned that she might pass out any moment or actually cough up a vital part of her anatomy. She pointed to her pocket and then grasped the rail again, her eyes even wilder than before. I fished around in her pocket expecting to find some kind of medicinal inhaler.

Instead I drew out a pack of Marlboroughs in a leather case with an attached lighter. The lighter had an outline of a woman on it like those I had seen on the back of truckers' mudflaps. I gave her a questioning look, and she took the bundle from me. She lit a cigarette and inhaled deeply. As she closed her eyes, a look of peace washed over her wrinkled face.

"I smoked a cigarette with the great Dian Fossey once," she said, looking down her nose at me with a squint against the smoke drifting up into her face. She turned to smile at the bushpigs, who had gone back to digging in the dirt. "Of course, that was before she got a machete in the head. I am from Sweden."

She took off walking again without giving me the chance to comment that it would have been difficult to smoke a cigarette with

Dian Fossey after she was dead, though with this woman, one could not be too sure.

We arrived at the gorilla shelter in a short time, and Karin pointed to a wooden bench close to the gorillas' enclosure. We sat down on it together. I looked around. The gorillas' enclosure was a generous quarter-acre with hills, trees, bushes, a stream leading to a pond, and all kinds of branches, sticks, and logs for the gorillas to manipulate.

The section of their habitat where we sat was covered by a large wooden grotto that was separated from the public by huge squares of heavy-duty safety glass that ran from the floor to the ceiling in a forty-foot half-circle. It was the nicest captive habitat that I had ever seen, and I had heard that Dr. Fossey had helped design it. It was probably then that she and Karin had smoked together. I imagined them both puffing away and silently surveying the enclosure with silent satisfaction between coughing fits.

"This is where you will sit and work. I brought out a little heater for you so you wouldn't freeze your ass off. The zoo's insurance doesn't cover ass freezage, you know!" She said the last sentence very loudly and looked like she was going to start laughing. Afraid that she would have another fit, I hurriedly asked her to tell me about the gorillas.

I learned that this group had four members: Adhama, the leading male, who was thirty-seven; Taufiki, a younger male who was eight and just reaching maturity; Binami, an older female of thirty-five years; and Nonesha, a five-year-old female.

Binami was pregnant with Adhama's third child and was due in July. Gorillas have a menstrual cycle in a monthly pattern much like human women, and their gestation period is almost identical.

"Binami got pregnant in October, then?" I asked.

"Yes. You should have seen it. When gorilla females come into estrus, or heat if you want to call it that, they start doing things they would never do otherwise. They walk right up to the lead male and stare at him about six inches from his face. Normally, this would be

Binami, Nonesha, Adhama, and Taufiki

a threat, because gorillas don't like to be stared at, you know. But Binami stared at Adhama without flinching 'til he got the message.

"They mated and were done in about two minutes. I guess he thought they were finished and went back to eating his celery. All of their matings before had been brief, usually only one or two copulations. But she turned around and stared at him again, just like before. He tried to turn away, but she stayed inches away from his face. They ended up mating thirty-three times that day. It was so funny, because he kept that celery in his hand the whole time and never got a chance to eat it. At the end of the day he came inside and passed out with that sorry wilted stalk still clenched in his fist."

I thought about the account I had read about two gorillas trying to mate on a hill and both of them inching farther down the hill as they went along until they finished at the bottom, both looking around, perplexed as to how they ended up there. Gorilla mating must be an all-consuming business.

Karin and I stood silent together for a few moments, and then she turned to me and told me that she had to let the gorillas out into the day habitat and then get to work fixing their lunches, a two-hour effort that involved boiling eggs, cooking yams and apples, chopping greens, and hiding vitamins in shredded-wheat biscuits. After she left, I sat alone and waited for the gorillas to come out of their night rooms, over the hill, and into their day.

Through the clouds of steam from my excited breaths I surveyed the habitat and waited for the gorillas. They had probably just finished their breakfast in the small set of connected rooms where they slept at night. These rooms had high platforms for the gorillas to sleep on, and they were filled with timothy hay for bedding. The gorillas' evening meals were also eaten in these rooms; while the gorillas were outside in their day habitat, the rooms were cleaned and the food was hidden in the hay and strung on firehoses that were slung like spiderwebs from wall to wall. These walls were painted with jungle murals of trees, vines, and animals, many of which had dubious claims to the rainforest background.

After breakfast—fed to them through sliding doors that opened between the night rooms and the large office and kitchen area—the gorillas were anxious to get out into the sprawling habitat and begin their hours among the grass and trees. Today was no exception.

Appearing over the hill came the gorillas in single file. As they crested the hill that separated the indoor gorilla complex from the habitat, I looked at each gorilla's file so that I could learn them by sight. Each had a thick dossier including individual photographs, birthplace, medical history, and notes on that gorilla's interactions with the group as well as comments on his or her personality. Karin had provided me with these files so that I might learn as much as I could about the gorillas in a short time. Eager to get a sense of the individuals I would be dealing with over the coming year, I read the files, looking at the gorillas described in those pages and hoping their histories would become as familiar to me as my own.

Adhama's Story

According to the dossier I held in my hand, Adhama was captured in the forests of Cameroon in February of 1953 and was estimated to be six months old at that time. I knew from reading about early capturing practices before the Endangered Species Act of 1973 that such captures were brutal. I looked up from my reading at his benign, placid face, and he met my gaze. I thought about his tiny eyes, still pale blue from birth, looking out on the world many years ago, searching the faces of his troop-mates as they perhaps heard the sound of the approaching hunters and turned in unison for a startled look in the direction of danger before the females swept up their young and ran away as fast as they could, leaving the male behind to fight for all of their lives.

Adhama would probably not have seen the death of the rest of his troop, though they were probably killed in the defense of their children. It is almost certain that his mother was shot from beneath him, making it easier for the hunters to pry him from her and stuff him screaming and flailing into the dirty burlap sack that was standard confinement for captured gorilla infants at the time.

Gorillas were called "black devils" by early hunters of the nineteenth century, and convictions of the day were built on the myth of the gorilla as a bloodthirsty man-killer: a myth built by white hunters seeking to make legendary their exploits in the "Dark Continent."[1] I remembered reading that hunters facing the grisly prospect of a confrontation with a ferocious male gorilla were advised to patiently stand their ground until the gorilla brought the gun barrel into its mouth, and then pull the trigger.[2]

Feeling vaguely embarrassed by such human subterfuge, I lowered my eyes from Adhama's gaze and back to the dossier once more. I learned that Adhama had arrived in London after a brief layover in Casablanca. He remained in his original crate until his arrival at the Harrington Zoo in Montana two weeks after his capture in the lush jungles of the tropical rainforests of his birth. He was dehydrated and suffering from a serious lung infection, and his constant diarrhea had not only made his crate unpleasant, but its filthy walls had caused his sensitive skin to erupt in sores. He almost died before he could be put on display. I remembered reading that for every baby gorilla who lived to be displayed, many times that number died during these early attempts to show them in zoos.

Adhama was one who lived. He recovered with the help of the zoo director and his wife, who actually brought him to their home and showed him the first tenderness he had received in weeks. The early records note that though he had chronic pulmonary problems, he retained a trusting and outgoing attitude throughout even the worst of the ordeal and soon became a favorite of the zoo staff.

Eventually, the Harrington Zoo was temporarily closed down for having inadequate enclosures for their specimens. Adhama's enclosure was among the worst. A concrete rectangle measuring ten by fifteen feet, the space had soon become completely inadequate; the zoo staff, who loved him, eased his boredom with hamburgers and chocolate shakes. When Adhama reached Whitman Pines on what is called "permanent loan," he weighed 600 pounds. He had never touched grass.

After a standard quarantine period of several weeks, Adhama experienced his spacious and verdant new home as mechanical steel doors slowly lifted to unveil what must have seemed a boundless world of green beyond the isolation chamber. A complete report was filled out by volunteer observers who had the joy of watching his first experience with a living habitat since he was a baby.

At first he could barely be seen timidly peeking around the edge of the doors, only to quickly back away as if the habitat beyond were on fire. Finally, he moved his body around to stand squarely in the doorway. He stared at the grass in front of the door for a full five minutes without even blinking. Then he put one hand out and moved it over the top of the grass with the pressure of butterfly wings. After moving his hand back and forth for a moment, he brought it to his nose and sniffed in tentatively. The smell of the green and growing must have stirred a long-forgotten memory, for an "ah-ha" look of recognition spread across his broad face, and he stepped fully out into the sunlight. He looked up at the sun with his eyes almost shut and let his mouth fall open. He grunted. He sat down.

Perhaps it did not occur to him to move any farther than the ten feet he had grown accustomed to, for it was days before he explored the full range of his habitat with its hills, stream, trees, and places to hide; but when he did, Karin had a hard time convincing him to come inside. To her credit, only once did she resort to the use of a chocolate shake.

Her notes in the dossier recorded that some nights he would stay outside, sitting on the top of the habitat's tallest hill, watching the stars and grunting to himself until he fell asleep on his back, having watched the stars until his eyes closed. She could not leave until he had come in for the night, and so at times she would go outside and sit on the hill with him. This was strictly against zoo policy, given his size and strength. She would often put on her tattered black coat to escape detection in the cool, earth-scented darkness, and after silently watching the sky with him for an hour, interrupted only by Adhama's occasional cough, she would get up, and he would follow her in.

She told me later that she knew he was lonely and felt that he was more willing to go inside if he could share his night watch with someone first. He was not to be lonely for long.

Binami's Story

Binami was captured when she was around eight months of age. The acquisition was undertaken only a year later than that of Adhama, but the notes briefly mention that the method of capture was somewhat different. In Binami's case she and her mother had been tracked for several days through the jungles of Gabon. When her mother made the mistake of moving beyond earshot of the troop, the Africans tracking them shot Binami's mother with a poisoned arrow. A description of this technique from the 1930s tells us that it was common for the hunters to follow the herd until a mother and baby were singled out, and then the mother was shot with a poisoned dart that slowly killed her. As she weakened, she fell behind the group, making it easy to take the infant from her body when she finally succumbed to the poison.[3]

This method of capture was felt to be a vast improvement for zoo animal collectors, as it was less dangerous for the hunting party, and direct confrontation with an angry male was not nearly as likely. However, the method did little to help the chances of survival for the newly captured offspring, and in Binami's case, the poison used on the fatal arrow had passed to her through her mother's breast milk, making her gravely ill.

Binami had the good fortune of being met in the nearest village by the future director of the Whitman Pines Zoo, who was then the director of a small zoo in Colorado. He was accompanied by the staff veterinarian, who specialized in primate care and had been brought along with the acquisition party to increase the captured specimen's chances at survival. The quick administration of atropine spared Binami any further ravages of the poison, but just as important, a woman from the village was hired to stay with Binami twenty-four hours a day and nurse the young gorilla from her own breast. The village children swarmed around the young woman as she sat outside the veterinarian's hut, staring at Binami attached to her breast sucking furtively for nourishment and comfort, as the young woman's own infant sucked at the other breast and gazed over at the tiny gorilla with interest.

After a few months of this around-the-clock care, Binami, the veterinarian, and the zoo director boarded the first plane of many that would take Binami away from that side of the world forever. The young African woman stood alone beside the crude dirt runway, staring with pained eyes at the plane that held her adopted daughter, and covered her mouth with her hand, fighting tears as the aircraft took off and grew smaller in the distance. Soon after, the director heard that she had been so brokenhearted that her husband and other men in the village organized a hunt to get another infant gorilla for her.

Binami lived at the Colorado zoo during her early years, and although she was alone in her enclosure for much of the day, in the evenings she was allowed to play with a group of three chimpanzees her age and a slightly older orangutan, six years old. She was unusually assertive for a young female gorilla; she was rambunctious, rough, and demanding, clearly the dominant member of her play group. The other young apes would generally give her any play items she wanted, to avoid the screaming and biting that ensued if she was denied her desires.

She was also quick to laugh, though, in her small, panting gorilla fashion, and showed concern if any of her playmates seemed to be hurt or in real distress. When she was seven she extracted an embedded splinter from one of the young chimpanzees' eyes, saving him the scheduled immobilization and operation planned by the zoo veterinarian.

When the zoo director was offered the post at Whitman Pines, his final contract stipulated that Binami would make the move with him so that he might accomplish his long-standing goal of starting a captive gorilla breeding project.

To this end, Binami met Adhama when she was eight and he was ten years old. Gorillas become sexually mature at these ages, female gorillas often achieving estrus cycles and fertility several years earlier than males. It was clear that Adhama was on his way to full capacity, for in the last year, the telltale whitish hairs of the mature "silverback" gorilla were spreading over his shoulders and beginning to creep to his legs.

Gorilla mating is a complex affair, built on rituals that sometimes last hours. The female will begin to give the male the "estrus gaze" that signals her desire to mate, but her readiness, especially with inexperienced females, is often punctuated by a discomfort with the unusually close proximity of the male; turning her relatively smaller frame to his massive body is a process during which she needs much reassurance. She knows that if the dance suffers a misstep, the male can become frustrated and attack her. Though females are very seldom hurt physically in these exchanges, the emotional harmony of the group suffers, and the pair have to start once more from the beginning for mating to occur.

There is evidence that much of this process is learned from watching older troop members. When gorillas do not learn these rituals through observation, it is difficult for them to ever really work it out. After several years of false starts between Adhama and Binami, the zoo staff was exasperated.

In desperation, someone suggested showing the two gorillas videos of gorilla courtship and mating when Binami was next in estrus. The day after the films, the gorillas actually performed the necessary steps, ending with Binami backing up into Adhama's lap. Sitting upright to avoid lowering his 550 pounds onto Binami's back, he began to copulate with Binami. Adhama gently patted Binami's back throughout the one and a half minutes it took for the mating to run its course.

They mated only this one time, but it was enough. I looked up from Binami's file as the result of this singular event now raised his head over the central hill of the habitat.

Taufiki's Story

Taufiki was born somewhat late in Adhama's and Binami's reproductive life. His birth was a major success; if he lived, the success would have far-reaching implications for the captive breeding program that the zoo had helped to spearhead in tandem with zoos all around the country.

His life, it was hoped, would be off to a happier start and in far more security than either of his parents had enjoyed as infants.

After many meetings it was decided that as the birth approached, Adhama would be kept on low doses of Valium to keep him calm, and the two expectant parents would not be separated. The birth was to be natural, and the baby would be taken to a nursery only if Binami showed signs of neglect.

A great deal about mothering behavior in gorillas is learned, much like courtship. In the early days of captive breeding, infants had to be taken from their mothers soon after birth as it became clear that the mothers were at a loss as to how to care for their young. As those first critical hours passed, early mothers would neglect to position the infants near the breast, and the infants would grow weaker and weaker. Soon the mothers would be anxious and constantly shaking the tiny infants to keep them conscious and alert.

Though many of these offspring did not hang on, many were taken to zoo nurseries and successfully raised in good health. They were usually introduced back into their troops, but it was known that their infants in turn would have to be taken from them at birth, perpetuating a frustrating and dangerous cycle.

The Whitman Pines staff, now including Karin Forrmen, hoped that this cycle would be avoided. Remembering that the courtship video had helped the gorillas with behavioral strategies, Karin borrowed videos of gorilla births from other zoos and showed them to both gorillas. She found that Adhama was not interested in them, though Binami watched them over and over. Early on, a stuffed toy gorilla was introduced for Binami to hold, and Karin practiced each morning with her, showing her how to hold the doll properly and bring it to rest on her chest. She also practiced getting Binami to bring the doll over to the square opening in the spare cage between the night rooms and the office. It was her hope that if the infant had to be taken from Binami, she could simply ask for it through the opening and spare both mother and infant the trauma of an anesthetizing dart. Such a procedure was not only traumatic, but risky; there was always a chance the dart could hit the baby. Karin hoped none of these measures would have to be taken.

Unfortunately, Taufiki's birth was not going to be easy. As

Binami's labor progressed it became clear that something was going wrong. Binami had been actively straining for quite some time, occasionally putting her hand between her legs and then sniffing her fingers as Adhama continued his vigil, stomach-down on the floor. As Binami sniffed her fingers again and again, Adhama followed the motion and searched his companion's worried face for some clue as to what was going on.

As still more time passed, the team of onlookers were starting to discuss what should be done. When Binami started to fall asleep between the close contractions, it became obvious that she was exhausted and a plan of action must be chosen quickly.

Karin went to the office refrigerator and, after retrieving Binami's favorite blue bowl from the shelf, began mixing her "surefire gorilla bait," consisting of tapioca baby food, sliced bananas, grape jelly, milk, and strawberries. After much coaxing, she was able to lure Binami into the spare night room, already prepared with hay and blankets. Quietly, from the other side of the hall, Adhama's room was closed off. Involved in eating his bribe of beloved celery, he paid no attention until Binami realized that her door, too, had been shut.

At her first piercing scream, he realized that he was helpless to defend her and gave a single mighty roar before beginning to bang the heavy metal door again and again with his enormous fist. This commotion from both sides was compounded when Binami immediately understood that the appearance of the zoo veterinarian was inevitable and added her screams to Adhama's.

The veterinarian was a long-time nemesis of the gorillas, who had no way of drawing direct associations about her miraculous cures; they equated her face only with a painful sting, a loss of control, and then blackness. Knowing that this was the case broke her heart, and she had gotten into the unconscious habit of saying she was sorry in a compulsive way during the darting process.

These were her words now as she positioned her four-foot-eleven-inch frame in front of Binami's window and pushed her long, red hair away from her face. Binami let out a threatening bark and shook her fist once in the air as the dart found its way from the small air pistol to target in her thigh. Though Binami reached down with lightning speed and pulled out the four-inch dart by the long red-and-yellow pom on its back end, the force of the impact had already injected the narcotic deep into the muscle of her upper leg.

Thankfully, though immobilization was always stressful for gorillas, new drugs brought unconsciousness quickly. The most widely used drug in recent years past was Ketamine, a drug that rendered animals unable to move but nightmarishly aware of everything that was happening around them in a distorted sense of reality, much like the twilight sleep generations of our foremothers had endured during their own labor and delivery experiences. Though most feeling creatures would understandably find the Ketamine experience at best unpleasant, it was, in fact, this very drug that hit the streets under the name "angel dust," where somehow it became associated with a good time. It is disturbing to think of all the zoo animals going back many generations who, in addition to being bored, lonely, and then sick, underwent the very worst of hallucinogenic "trips" on top of it all.

Mercifully, in Binami's case, her pain was gone, and she snored softly and peacefully in her birthing corner at the end of five minutes. Those swinging into action around her might have wished they, too, had this luxury as Adhama's screams continued to pierce the air, punctuated by the deafening banging of his fist on the dividing door only a few feet away from them.

Karin, the veterinarian, and two volunteer birth observers crouched over Binami's now prone form. The veterinarian put her small hand into Binami's vagina and then bunched her brows in an exasperated expression as she dropped her head in anticipation of the possible outcomes.

The umbilical cord was wrapped double around the baby's neck. One loop she was able to gently work around the top of the baby's head. The other was tighter and was obviously going to need to be cut. Inching the baby out, she poised the scissors and tried to brace herself against the screaming and the banging, occurring in irregular intervals for which it was impossible to prepare. With a deft clip during a pause between the worst screams and reverberating blows, the cord was severed, and the veterinarian immediately wrapped the baby boy in a receiving blanket and then sped him to the nursery and into a waiting incubator flooded with rich oxygen.

While Karin stayed with Binami and ensured that she slept off the drug peacefully, a small and helpless Taufiki battled for his life in a building on the other side of the zoo grounds. As days passed, it was clear that he had suffered a stroke during the labor, and his right arm and hand were useless. Like his father, though, he seemed naturally bright and tenacious in the face of spirit-bruising challenges. He soon began reaching up with his good left arm to anyone passing by his incubator, his crib, and finally his playpen.

From the earliest time, Taufiki had a peculiar squeaking laugh much like that of a human infant. He was so sociable that the zoo administration had agreed to open a special window to the nursery so that the public could see this long-anticipated tiny ambassador. A nursery attendant would rock Taufiki and nurse him while he smiled and laughed at the visitors streaming by outside his window. Sometimes he would imitate the faces they made at him and delight the visitors with his wide-eyed antics.

Gorilla babies are about four pounds when they are born, and they are about six months old before they gain the strength and coordination to start moving about. They are dependent on their mothers both in captivity and in the wild for up to five years, often nursing until this late age, though by the end of this time nursing is more for comfort than nutrition, the babies having begun to eat solid foods at about a year of age.

This period of dependence is a pivotal one for baby gorillas. During this period they learn to bond, first with their mothers and then with the others in their group, thus laying the solid foundations for a grasp of the social rules they will follow for the rest of their lives. The zoo staff was committed to returning Taufiki to his parents as soon as possible, and six months to the day after his birth, he was successfully returned to his mother after weeks of hourly visits from the nursery to Binami and back again.

Enlisting the old doll that she had used to help Binami practice for the birth (though it was a little worse for wear now), Karin taught Binami how to pay special attention to Taufiki's right arm and hand, to groom and massage them, and to tuck them in securely rather than let them dangle limply.

As time passed and he was a year old, there was some concern about how he would cope with the crawling stage of his development. To everyone's amazement, Taufiki never crawled. He learned to walk completely upright at all times and continues to do so even now that he is eight years old and has regained the use of his arm.

Nonesha's Story

Nonesha's birth took place in the same kind of serenity that had been hoped for during Taufiki's birth years earlier. The technological blessing of an ultrasound two days before had shown that there was every chance that this birth would be free of complications, and so it happened.

Adhama again stayed away from the room where Binami was beginning to show the early signs of labor, though, as before, he was given free access to be near her. Male gorillas in the wild also follow this pattern of allowing birthing mothers both physical space and protection; mothers need both forms of support to feel comfortable through the process. Taufiki, however, followed his mother from the nesting platform to the floor, intently watching her attempts to ease her discomfort. Taufiki had never seen a birth and was anxious about his mother's grunts and repeated movements around the room.

Binami, for her part, must have remembered the experience

of labor and birth, for her expression was one of quiet resignation rather than bewilderment. She did not make several nests this time but after several hours began to construct a single one in the corner where Taufiki's birth had taken place. Though this nest was not as sturdy as her regular constructions, it provided a soft place that seemed to ease her increasing pain and, to the relief of those keeping vigil, would offer a cushioned place on the floor for the baby's arrival into this world of hard edges.

True to the patterns observed in the wild, it was as the night deepened that Binami came closer and closer to giving birth. In the wild, gorillas are stationary at this time, each troop member having constructed his or her own nest and not likely to move about until morning.

At around 3:00 A.M. Binami sat upright in the corner of the room she had used during the first birth. She placed her legs far apart and rested them on the floor. All at once a gush of liquid, baby, and blood issued from between her legs. Binami caught the baby and, after looking completely surprised for a moment, began to eat some of the afterbirth as she wiped Nonesha's tiny closed eyes and nose with her large, calloused hands. It has long been believed that antibiotic properties in the placenta can benefit the mother and offspring when ingested following a healthy birth. Gorillas in the wild will not eat the placenta of a stillbirth. That Binami was eating bits of placenta and wiping the baby's face were both good signs.

After some time, Binami gently shook the infant. Nonesha opened her blue newborn eyes and gazed at her mother, letting out a wide yawn and stretching her flailing skinny arms. Binami put the five-pound infant to her chest, and Nonesha rooted around for a nipple, closing her eyes as she latched on. Binami rumbled a grunt of satisfaction and fell asleep still sitting in the corner, never once letting Nonesha far from her chest and never shaking her again.

Though the door to Binami's night room had been left open, Adhama was content to sit in the doorway and watch the whole process silently. Now that Binami was asleep, Adhama poked his massive head through the door and stared at his daughter. He too belched a satisfied grunt of contentment and settled with his chin in his hands to watch his sleeping family for the rest of the night. Occasionally, he looked over at Karin, who was also now asleep in her chair.

It was obvious from an early stage that Nonesha's personality was like that of her mother. When she could crawl, she would scoot purposefully over to her father, who towered over her and weighed 530 pounds more than she did. Too young to eat solid food, she would nevertheless try to reach for the food items he had in his hands. If he didn't indulge her by dropping the item (which he usually didn't) she would scream and bunch up her uncoordinated muscles in such a way that led to her falling on her face. He would sigh deeply and pat her softly on the back, at which point Binami would rush over to whisk her away, looking apologetic.

By ignoring her histrionics, Adhama helped Nonesha grow up to be more compromising and patient than her mother. Her days of safety and freedom from hunger and cold put a certain care-free spin on her personality. Her emotions were always near the surface; she was given to exuberant dances and displays of somer-saulting. Though many gorillas are leery of water, Nonesha often waded out into the pond, going far enough that her chin would brush the water, much to her mother's consternation. Binami would furiously pace the shore, issuing sharp grunts of admonishment until Nonesha climbed out to rest on the dry ground; gorilla mothers seem to know that, because of even a young gorilla's muscle mass, their children can sink and drown. And so, under her mother's vigi-lant eye, Nonesha grew through a healthy childhood.

Gorillas show a strong aversion to incest, and although Nonesha's fertility would become evident as she grew older, both she and her father would never consider each other as suitable mates. This future dilemma was far from Nonesha's thoughts at the moment. She seemed to relish the fact that she had few responsi-bilities, being the sole child with the full attention of two adults, something unusual indeed for gorilla offspring. But zoos around the country had turned their thoughts toward Nonesha's future. The captive breeding project was doing well now, and gorillas were routinely sent to other zoos to start breeding families. Though females were hard to come by, the zoos expressing an interest in Nonesha could not provide a natural habitat for her. The Whitman Pines Zoo director fought to keep her with her family rather than send her to live in a concrete box, though he worried that she would become socially bored in her current situation. Soon, however, Binami was pregnant again, and new life would again be breathed into the small gorilla family sometime during the summer of this year.

I put down the dossiers, mulling the stories over in my mind. After all of the gorillas came over the hill they made for the sheltering roof of the grotto and began looking for celery and yams hidden around the enclosure at breakfast. I watched them for a long time, scribbling some notes on the data sheets I had constructed to chart their behavior. I had almost become convinced that they took no notice of me at all when I realized that they were each stealing sidelong glances at me whenever I seemed to look away.

After about twenty minutes, Adhama approached the window cautiously, pretending to be interested in a beetle that had landed on the window. He got about two inches from the bug, staring cross-eyed at it. At shorter and shorter intervals, he looked over to where I was sitting and studied me from head to toe. He came over to sit directly in front of me and leaned forward to stare me in the eye for a moment. It was a piercing look, without compromise. In ways it reminded me of the look Karin had given me earlier, during our first meeting, but it also had a different quality. It seemed to say "I know who I am and I know who you are. Do you know who you are?" It was a look that could only have its genesis in captivity. It was the look of a soul and mind that had accepted the reality of their bondage and from this understanding know the limitations of all other things.

Sitting with Adhama

I knew from my readings on gorilla behavior that looking away would have been the accepted thing to do, as apes often interpret a direct stare as a threat. Then I remembered Jane Goodall saying once that if you have nothing to hide, a kind stare into the eyes of an ape you've just met usually isn't taken the wrong way. Well, we all have something to hide, perhaps especially the nagging feeling that as a species we've given some pretty raw deals to other living things; but I did my best to meet his gaze, from a foot away, with a strength and respect I hoped were genuine enough.

Slowly, he leaned even closer until his forehead was touching the glass. I leaned forward, too. We sat like that for a while, forehead to forehead, still eye to eye.

I wondered why I didn't feel elated. For some reason it ran

through my mind that I could get up and walk out of the zoo and never come back. Then, without the expected feeling of triumph in making a transcendent connection with a gorilla, but with a melancholy that would become familiar, I made the realization that I could never leave again. With an understanding that was a long time rising to the surface as I continued to look into Adhama's eyes, I knew that life itself can never either be held captive or be really liberated. I knew that we are all prisoners on either side of the glass, free to decide whether to live a life or live a death. I knew that like the gorillas, I had decided to live a life. From now on my life would be bound with that of the gorilla nation.

Book of Days

February 28 I sat in my usual place on the wide oak bench and watched the gorillas coming out in single file. Some early visitors were milling around in front of the grotto window, patiently waiting for the gorillas to appear. An excited buzz went up when Adhama's massive head peeked over the hill at the far end of the enclosure. Parents pointed and told their children to watch.

Murmurs of astonishment drifted back and forth as the gorillas made their way slowly across the habitat and approached the window. People were never prepared for the gorillas' size and presence. Parents began counting the gorillas with their children: "One . . . two . . . three . . . . Wait a minute, what is that at the very back? Is that a *man*?"

I was to hear this last question many times as the distant and blurry image of Taufiki, walking upright, came closer to the onlookers.

Though Taufiki's standing and walking upright always amazed the visitors, I had begun to realize after watching the gorillas that they frequently stood upright and walked. Standing to reach food was extremely common. The gorilla enclosure was dotted with trees, tall bushes, and high rock walls, all of which served as hiding places for food and browse. Hiding the gorillas' food in these high places beyond reach always led to their standing and foraging during the day.

Twice I had seen Nonesha stand up and carry tools from the place she found to a separate workplace some distance away. Once she carried two tools to a rock wall to use them for an activity, and another time she carried a long stick down to the pond (walking a distance of ten meters) to reach a candy wrapper floating there. More routinely the gorillas walked on two legs to carry food, taking it back to the shelter and their nests.

Taufiki and Nonesha enjoying a break from wrestling

Standing up to look over vegetation was also something the gorillas did often. They would look over bushes to find their family members, and standing to peer over bushes was a routine part of Nonesha and Taufiki's games of "hide and seek." They took turns hiding or scouting around the bushes waiting for the perfect time to spring on each other, after which rousing bouts of tickle-biting ensued. They would both laugh their panting gorilla laughs with their lips sucked over their teeth in unique gorilla smiles until they were exhausted and wet with spit from each other's tickling teeth. Eventually they would lie in the hay helplessly, trying to catch their breath, until one would cast a sidelong glance at the other and then leap with renewed vigor, starting the whole thing over again.

Wild gorillas have also been seen to stand upright, or "bipedally," to reach for food.[1] The most common context for bipedalism in wild populations of gorillas is during ritual/charging displays, when males are either challenging individuals or groups within or outside their own groups, or defending their groups.[2] Because these displays entail running upright with branches or dirt clods that are thrown at the end of the charge, it could be said that the gorillas are also technically transporting tools.

Dian Fossey related with humor the instances of young gorillas carrying what could be considered toys, in the form of *mtanga-tanga* fruits, standing up with the stalks in their teeth to beat the fruit against their chest.[3] During these occasions the gorillas were more interested in the fact that they were playing than the fact that they were carrying food items or toys bipedally. Young gorillas often take upright postures during play and sometimes run and stand to slap at their play partners; Fossey describes being slapped quite vigorously in such a playful bout!

Pioneer gorilla researcher George Schaller recorded gorillas standing upright to look over dense vegetation.[4] He also made the observation that gorillas will walk on two feet to shelter during rainstorms to avoid getting their hands and chest wet.[5]

Walking upright, in the minds of anthropologists and laypeople alike, has traditionally been one of several strong dividing lines between humans and apes. All of these dividing lines have been erased and redrawn in the sand as scientists—and everyone else—look for the elusive lines that will succeed in keeping us divided from the animal kingdom. Some criteria for division have been finally tossed out, though in their day they seemed as defensible as the dubious lines of walking upright, language, abstract thinking, and emotional life do today.

In 1904, riding the crest of Darwinism and eager to cast their vote for who was human and who was not, the Americans of the World's Fair in St. Louis put tribal peoples on display in exhibits supposedly reflecting their "natural habitat." These "habitats" were in fact woefully inadequate to protect the inhabitants from cold, hunger, and boredom, much like "naturalistic" zoo habitats today. One of the real goals of the habitats was to provide what amounted to crude factories where the indigenous people could produce authentic tools and weapons for sale. The great Apache leader Geronimo was on display there, selling as trinkets the arrowheads he made.

But of all the exhibits, the Congo Pygmies were the most talked about, as no one was sure they should be considered human. To encourage serious debate on the issue, officials organizing the fair had made a shopping list of human specimens for Samuel Phillips Vernor, the person put in charge of completing the World's Fair expedition into Africa for that purpose. The fair committee wanted twelve Pygmies, including a patriarch or chief, an adult female, two infants, and an assortment of young and old Pygmies.[6]

A man named Ota Benga was one of the Pygmies (or Mbuti, as they are now called) who eventually came to be displayed at the World's Fair. He had been away from his village hunting forest elephant when the village was wiped out by marauding forces put into action by Belgium's King Leopold, who controlled the Congo and was devastating vast populations of people and animals to satisfy his lust for ivory. His forces heard that the Mbuti had stashes of ivory and massacred entire villages to get to it. Coming back from hunting, Ota had found his entire family and village killed or wounded and Leopold's forces still on the scene. Though he fought them, the party captured Ota and bound him. Realizing that his life as he had known it had ended, there was no reason for him to fight when Vernor, looking for a bargain, bought him from Leopold's Force Publique, and so Ota soon found himself on a boat destined for America to be shown at the fair.

The weather was cold in St. Louis, but the fair committee felt that giving the Africans warm clothes and blankets would not enhance the realism of the habitat they wanted to display them in, and so the Africans shivered in their loincloths and huddled in the flimsy shacks in their enclosure.

Once on display, Ota was the subject of many scientific tests. Was he able to discern the color blue? How did he compare with defective Caucasians on intelligence tests? How quickly would he respond to pain?

After being measured and tested in every conceivable way until the fair came to a close, Ota ended up a regular exhibit at the Bronx Zoo, where space was made available for him in the ape and monkey house. People would watch him interacting with the other primates and continue to ask if he was a human or an ape.

Eventually crowds tired of Ota, and he was left without even the mean resources of the zoo or the scientific community. On March 20, 1916, wanting to go home to Africa but having no money and no one interested in taking him back, Ota built a fire at five o'clock in the afternoon. He broke off the caps that had been placed on his filed teeth to make him look less fierce. He took off all his clothes and began to dance. Still dancing and singing, he took out a pistol and shot himself in the heart.

I thought about the questions we ask about the apes in our zoos and in our studies, questions often designed to shore up the wall between us: Can they discern colors? How does their intelligence measure up to a human's? How long does it take them to react to pain? We avoid asking if they long for home or acknowledging that whether or not they are human, they deserve agency. If a gorilla can walk up to the zoo gates on two legs, should he be able to keep walking? Ota Benga was not allowed to.

March 1 Taufiki came over the hill, a dark silhouette swaying back and forth with his swaggering gait. Because gorillas' legs come down in a straight line from their pelvis to the ground, they must shift their weight back and forth over each foot as they take steps, unlike humans, whose legs come down at an angle to a point under their center of gravity with each stride.

I was already becoming fond of Taufiki. He was good-natured and gentle, amiable with his family, even though I knew his unusual posture must cause him chronic pain and perhaps even headaches. In humans the foramen magnum, the hole in the skull where the spinal cord passes from the brain into the spinal column, is almost directly below the skull. In gorillas this hole is located more toward the back of the skull, a more comfortable place allowing the gorillas to walk with their head up and facing forward while they walk along on their feet and the knuckles of their hands. Because Taufiki walked upright all the time, he had to keep his head tilted down at an unnatural angle. The pressures on his neck, back, and head muscles could cause not only headaches, but difficulties in chewing, swallowing, and vocalizing.

If these challenges were wearing on him, it was not showing. He reached high into the bushes and, with the tips of his fingers, worked a clump of apple peels into grabbing range. He brought the delicacy down to his nose, sniffed it, and touched it with his tongue. He turned and gave the sticky mass to Binami, who had been standing behind him, watching his operation with keen interest. She took the sweet mass from his hands and put the entire wad in her mouth after brushing away a bee that had landed in the middle of the pile. She chewed it slowly, making rumbling noises of satisfaction. Taufiki turned to find peels for himself after he was sure that his mother was enjoying his offering. Once again he reached high into the bushes, at one point balancing on the toes of one foot.

Adhama had come to sit beside me, and now we were divided only by the window between our shoulders as we watched Taufiki's

Taufiki stands up in the trees to look for his family

impressive feats and Binami's riveted attention to her son's acquisition of the prized apple peels.

Adhama and I would watch for a while, and when I would stop to scribble notes on my data sheet, he would turn and look down his nose at the end of my scratching pen. Sometimes I would hold the page up for him to look at. He would study the chaotic lines and then turn his attention back to his family with a soft grunt.

He watched Nonesha come across the habitat with two fists full of romaine lettuce, celery, and apple peels and two large yams stuffed in her mouth. Intermittently she would stand up to adjust the yams and walk a few steps in this way. She finally made it to her nest in the grotto and sat on its edge while she arranged her booty in a neat row.

As I watched the gorillas pausing here and there to stand and reach bunches of romaine lettuce and ribbons of apple peel thrown high into the bushes along the path, my thoughts returned to walking upright. I thought again of the list of behaviors that we as scientists and humans have made to separate ourselves from the gorillas I now watched. It was as if each behavior was somehow an entity within itself and not part of a continuum of behaviors bleeding into form, thought, history, and environment.

Even college-level anthropology books are firm about the fact, for instance, that the human ancestor *Homo habilis* began using tools (we know that because we have found stone tools associated with their fossil remains) and that this tool-making ability placed us firmly on the path to civilization. In somewhat circular logic, the textbooks explain that humans could use tools because they were standing and walking upright, and the reason they were standing upright is because they needed to use tools. Though it is true that these feedback loops do reinforce each other, science is often silent on the subject of the long periods of trial and error involved in the acquisition of behaviors.

Perhaps because the development of tool use is untidy, the textbooks still rarely talk about the gradual experimentation with various tools fashioned of wood or other perishable material available to our ancestors millions of years before *Homo habilis*, who lived only two million years ago. There is little exploration of the millions of years of intermediate stages that do not happen to have pictorial reconstruction.

We have all seen the graphs: clear and straight black lines cutting boldly through the neatly delineated epochs dating back 65 million years. Often these graphs are superimposed with skulls representative of stages in our ancestry. "At this point *Australopithecus* stood up and walked." "At this point *Homo habilis* mastered the manufacture and use of tools." "This is when humans harnessed fire." And so on. We get the impression that one sunny day, every *Australopithecus* suddenly fell dead on his or her face as, phoenix-like, *Homo habilis* rose from their ashes, having absorbed all of their technological advances. The process continued, presumably until we, as finished human beings, made a final and triumphant step out of the ashes of our past, the real purpose of the whole show.

In reality an infinite number of divisions arose between these so-called individual species, and if we were to see them all lined up in progression, it would be impossible to draw a line anywhere. As it is with form, so it is with behavior. Things are tried and abandoned only to be tried again and repeated. Behaviors catch on very rapidly and then disappear again. One group leaves a behavior behind, and it is adopted by a different group. No line of progression in behavior stretches unbroken into the future, leading forever forward to a simple definite result. Everything occurs in context.

In 1981, John Gribbin and Jeremy Cherfas advanced a theory that respected this chaos and contextuality as it applies to bipedal walking. Basing their hypothesis on the fact that no fossils have been found that are ancestral to gorillas and chimpanzees but are not ancestral to humans, Gribbin and Cherfas believe that our common ancestor walked upright.[7] These scientists believe that early bipedality was a result of the spreading grasslands of the Old World about 20 million years ago and that for several million years following, while the grasslands stretched over Africa and Eurasia, our common ancestor took advantage of upright posture to look over tall grasses and wade through seas of grass between pockets of trees.

When the climate changed about 6 million years ago, say Gribbin and Cherfas, some members of our common ancestral form once again began to take advantage of a growing forest habitat and adopted a four-footed knuckle-walking mode of locomotion. This branch eventually gave rise to gorillas and chimpanzees. The branch that stayed in the grassland habitat eventually gave rise to human beings.

There is little doubt that tool and weapon use could develop into a more successful strategy in the grasslands than the forest: food was harder to procure without tools, and without weapons protohumans would have been more vulnerable to predators. Also, being able to carry food and tools to safe places to process and eat foodstuffs might naturally evolve into the discovery of containers. Thus the crude evolving material culture of these protohumans intertwined with walking upright and the surrounding habitat. The same could be said for gorillas' locomotion, material culture, and environment, all blending together without any real dividing lines.

March 2   As it might have been in the grasslands of the early savannas, it was bright and sunny today, and the gorillas seemed to be in high spirits. Perhaps because of the warming weather after a particularly bitter winter, Nonesha was in the mood to experiment; whatever the reason, I saw something extraordinary happen soon after the gorillas came out across the yard and into the grotto.

While the other gorillas began sifting through the hay and among the bushes for their morning delicacies, Nonesha, in uncustomary gregariousness, came over to the window where I was sitting and taking the first notes of the day. I put my hand on the glass, and she reciprocated by placing her larger hand opposite mine. I smiled and was pleased by a gorilla smile in return.

We sat amiably, both enjoying the warmth new to this year. Several minutes went by. I realized that Nonesha was staring at the rock wall of the grotto with intense interest. A moth high up on the wall was flying and then landing in the same place again and again. Nonesha stood up and approached the rock face. She tried to swat the moth, but it was too high for her to reach. She sat down facing the wall and stared at the moth for a long time.

Finally, she turned from the wall, and I thought she had grown bored with the moth. My curiosity was piqued, though, when she began pushing the hay in the grotto aside, clearly looking for something. She found a short sturdy stick about two feet long and tucked it under her arm. She continued to push around in the hay until she found a second longer, thinner stick. This search lasted a full minute, while a look of almost painful concentration knotted Nonesha's face.

Returning to the wall and checking to see where the moth was, she carefully placed the short stick at an angle against the wall directly below the moth. She gingerly balanced herself as she stepped onto this short stick, gaining about a foot and a half in height. Keeping the moth in her sights, she swung the second, thinner stick at the moth, hitting it dead center.

The moth fell to the ground. Nonesha immediately dropped her weapon and jumped down from the ladder stick, deftly snatching the wounded moth from the hay and popping it into her mouth. She let out deep grunts of satisfaction as she mashed the moth around her mouth and over her tongue, savoring what she must have considered a rare treat.

It was astounding. Not only had I never heard of gorillas using two tools in combination, but seeing a gorilla actually hunt with a weapon was something new indeed. Dian Fossey had recorded mountain gorillas eating grubs, worms, and snails, often ignoring other favored foods to hunt for these animal foodstuffs. When worms were found, they were immediately torn in half and then each half was eaten separately. Realizing that gorillas actually hunted sources of animal protein, Fossey eventually began to include boiled hamburger in the diets of the orphan gorillas she nursed to health. When she began offering the hamburger, she reported that the gorillas would eat it before anything else.

Since Dian Fossey's observations, other researchers have made careful studies of the eating of insects by gorillas.[8] One of the most frequent questions I've been asked about the gorillas has been what they eat. Most visitors assume, as did many early scientists, that the gorillas are complete vegetarians. Some gorilla ethologists later began to factor insect protein into wild gorilla diets, reasoning that with the huge handfuls of vegetation they place in their mouth, gorillas must inadvertently ingest some unsuspecting fauna. Though in many cases swallowed unintentionally, these tiny packets of concentrated protein must have an impact on the gorillas' dietary needs.

In the earlier days of captivity, gorillas were routinely fed meat but, like modern humans, probably got too much of it. When popular consensus turned toward the picture of the strictly vegetarian gorilla, many zoos reflected this in their gorillas' menus. Now that it has been documented that gorillas actively seek out animal protein, it is not uncommon to see gorillas once again offered meat, though many zoos meet their protein requirements with dairy products.

Of course, it has been demonstrated that the bulk of the gorillas' diets in the wild comprises fruits and vegetables. The prodigious amounts of fibrous fruits and vegetables that the gorillas eat are responsible for their potbellied appearance, as foods must remain in the stomach for long periods of breaking down and digestion. Zoo gorillas are no exception. In captivity, a typical daily gorilla diet might include two pounds of oranges, eight pounds of bananas, four pounds of carrots, one-half pound of potatoes, five pounds of apples, three pounds of celery, three pounds of lettuce or greens, two pounds of tomatoes, and perhaps another six pounds of miscellaneous fruits or vegetables in season, as well as a pound of meat or an equal amount of yogurt or milk.

Vitamin and mineral supplements have also come into use, and many years ago the gorillas at Whitman Pines began enjoying a blend of bananas and yogurt, along with vitamins E and B complex, antioxidants, and calcium each morning. This dietary regimen puts the amount of food consumed by captive gorillas at about five percent of their own body weight each day.[9]

Though captive gorillas' zoo diets are fairly homogenous throughout the world, strong evidence suggests that gorillas, like chimpanzees, have "food cultures" particular to their different regions. Gorillas in the Ndoki National Park of Congo were found to eat the fruits of the *Polyalthia sauveolens*, for example, whereas the gorillas of Lope Reserve in Gabon did not. The Lope gorillas did eat the leaves and bark of *Milicia excelsa*, though the Ndoki gorillas avoided it.[10] These preferences seem to be based exclusively on acquired tastes: habits passed from one generation to the next. By way of these habits or preferences, which are the essence of culture, new behaviors such as Nonesha's taste for moths and hunting them with weapons are passed down and become indispensable to later generations.

It is interesting to consider how many brilliant moments like Nonesha's now occur out of natural context and without the continuity of generations. These experimental behaviors will likely not become part of a larger gorilla culture. Or, maybe, there is a gorilla collective unconscious, and we will be surprised over the years to see emergent patterns of behavior sweep captive populations of gorillas. . . .

March 3 I arrived late at the zoo amid a downpour of warming spring rain. As I took my seat on the wooden bench and pushed the dripping hood back off my head, I looked out into the habitat and realized that the gorillas were clustered around the rock wall toward the back of the enclosure where Karin stood high above them to toss snacks into the habitat each midday. She had just begun to pitch down the coveted treats from the shiny silver bucket beside her feet, and the gorillas' faces were all turned upward expectantly, blinking against the heavy drops of rain.

She caught my eye and waved to me with one hand while stopping to slick back her soaked hair with the other. She was smiling, and the warmth so characteristic in her face contrasted with the drenched green foliage around her and the chilling gray of the sky.

The gorillas turned briefly to regard me, then quickly returned their attention to the bucket.

Apples (always baked, and sprinkled with cinnamon on in-clement days), then oranges, shredded-wheat biscuits, celery, and peanuts came down with careful aim to each gorilla in turn, always starting with Adhama. Most times the gorillas actually caught the items as they came arcing down.

When the bucket was empty, Karin sat down and dangled her legs over the wall, talking gently to the gorillas as they polished off the leftovers in the grass around them. She seemed to be waiting for something.

Before long, Adhama came closer to the wall and sat down once more. Fixing Karin with an intense stare, he raised his arm toward her and then moved his outstretched hand repeatedly in a motion that was, in American Sign Language, the gesture for "give me, please." Karin's smile grew broader as she fished around in the pocket of her torn black coat for the prize Adhama knew was there.

With a flourish Karin withdrew a boiled egg and deftly tossed it to the waiting silverback. He caught it in his mouth and, to my amazement, gently cracked it with his teeth and peeled it using only his lips. He bit down, and a tiny trickle of soft yellow yolk spilled from his lip and ran down his chin as he closed his eyes in rapture. When he had swallowed the thoroughly masticated egg, he grunted happily and clapped his hands together loudly.

Karin then produced three more eggs from her pocket, tossing one each to Binami, Nonesha, and Taufiki in turn. Adhama repeated his begging gesture, but Karin turned her coat pockets inside out to prove there were no more eggs. In disgust, he shook his fist at her and made a raspberry as he turned away, moving toward the grotto, as the fun was clearly over.

Unruffled by his change of regard, Karin cheerfully waved to me again as she turned and walked away down the trail that led from the wall-top to the office area. She was whistling and swinging the bucket in an exaggerated way as she disappeared from sight.

The gorillas came into the grotto in single file, and for the first time, each one came by my place at the window to say hello. Binami was first, putting her face close to the window and looking down at my notes. It would become a habit each morning for me to show her what I was writing, then take apart my pen to show its guts while she watched in fascination, not diverting her eyes until I put the pen back together to write another line.

Nonesha came by next and pressed her lips against the glass. I kissed her back from the other side, both of our eyes crossing as we tried to continue looking at each other. Taufiki pushed her out of the way and playfully threw hay at the window, doing a funny little hopping dance that ended with him biting his own wrist and then falling down and kicking his feet furiously in the air. When he sat up and looked for my reaction, pieces of alfalfa clung at wild angles to the top of his head. I laughed. At this, he started the whole routine over again. By the end of it, though, he was bored and left to find something else to do.

Adhama came to greet me last, which was unusual. His brow was furrowed as he sat directly in front of me. Then he made the asking gesture to me. I was unsure of what to do. He made it again. I pulled the lunch sack out of my backpack and showed him the contents: some leftover macaroni and cheese, a nectarine, a plum, some carrots, and a bottle of Gatorade.

He pointed to the Gatorade. By pushing the bottle against the window and shrugging my shoulders I tried to explain to him that there was no way to give it to him. He pointed to the wall and raised his eyebrows. "Throw it over. . . . What kind of stupid gorilla are you?" he seemed to say. I shook my head and pointed to my seat and notes, in a feeble attempt at demonstrating my duties. He banged the window with his big, hairy fist, let out another raspberry, and turned his back to me. Occasionally he would turn to look over his shoulder and purse his lips at me. He did not need to say it in English; I had no doubt what was going through his mind.

The tantalizing frontier of apes' language capabilities has

drawn interest and controversy. Most people would concede that apes, being highly social creatures, communicate with one another in very complex ways in the wild.

"Hoots" emitted by gorilla males warn of impending threat charges, and unmistakable "wrraahs" are the deafening roars that males project when they encounter outside species that may endanger the well-being of their groups. Both "hoots" and "wrraahs" are used in the same ways by gorillas and chimpanzees, and some researchers contend that this is strong evidence that their common ancestor also used these calls.[11] Gorillas make long rumbling grunts or belches in their throats when they are happy and content. When feeding in the wild, gorillas use these contentment grunts to maintain contact with their group-mates in dense vegetation and perhaps to reassure one another that there is no danger at hand. Gorillas cry when they are sad. Their crying sounds like human crying, and though they are not known to shed actual tears, Dian Fossey claimed to have seen a gorilla cry tears after being captured and taken away from her family. Fossey also made recordings of gorilla vocalizations accompanying their physical signals.[12] Within these rich contexts of communication gorillas laugh and smile, frown and show surprise.

Adhama's use of signs we both understood stemmed from a rich repertoire of gorilla gestures: head-bobbing as a mild threat, crouching down as a means of avoiding a fight, and pats, hugs, and kisses given as reassurance. Even the gorilla gesture of chest-beating, so caricatured in popular fantasy, is complex. Accomplished by hitting the chest, stomach, or thighs with cupped hands (not bunched fists), the rapid "pukka-pukka-pukka" sound can denote agitation, joy, exuberance, an invitation to play, or an assertion of dominance. Probably much of the subtlety of gorilla body language and its rich vocabulary are lost on us, the human word-junkies. It is perhaps no surprise that apes have learned to communicate using human language but humans have made little progress in returning the favor.

Language studies with chimpanzees and bonobos (often called

pygmy chimpanzees) have led to surprising results, showing levels of language comprehension beyond anyone's expectations, to the applause or consternation of those on either side of the fence of human/ape separation. Though many people have heard of chimpanzee language studies, fewer have heard of the language studies with gorillas (with the exception of Koko). In learning complex tasks, gorillas have long had the reputation of being slow and obstinate.

Dr. Francine "Penny" Patterson is one gorilla language researcher who believes that gorillas' stubbornness and asserted independence is actually a sign of their high intelligence and creative thinking styles. Dr. Patterson's studies of gorilla language capabilities started in 1972, while she was a graduate student at Stanford University. Her first pupil was a lowland gorilla named Koko, then a very young gorilla who had been born in captivity. The two spent hours together each day, Penny teaching her small companion the rudiments of American Sign Language.

The project has continued; currently Koko has a vocabulary of around a thousand words and can arrange these words into sentences averaging three to six words. She has begun to learn written English and recognizes many written words, including her name. She understands spoken English as well, and her receptive vocabulary is several times that of the number of signs she regularly uses. Using these skills, she has scored a 95 on the Stanford-Binet Intelligence Test.[13]

She uses her language skills in humorous ways as well. "That red," she signed one day to a staff researcher, indicating a nearby towel. The researcher, puzzled, told Koko that the towel was white, not red. "That RED!" signed Koko. These exchanges continued until Koko picked a tiny piece of red lint off the towel and held it up to the researcher. Grinning, she signed, "That RED!" She was playing a joke.

Koko also engages in imaginative play, both alone and with others; understands the concepts of past, present, and future; and in a combination of these abilities, anticipates the reactions of others and will lie to avoid consequences. She understands that work and rocks are both hard. She has described the color brown as also being hard, the color red as warm, and violet-blue as sad. She has used these creative abstractions to make up words, too, using "alligator milk" for milk that has gone bad (she hates alligators), a "bottle match" for a cigarette lighter, and "eye hat" for a mask.

This skill in abstraction has afforded us insights into the emotional life of gorillas. Koko can share, for instance, her reflections on death. When asked why gorillas die, she signs, "Trouble, old." When asked where gorillas go when they die, she signs, "Comfortable hole bye." She becomes agitated when people try to talk to her about her own death or the death of those she loves. She has lost many pets, including a cat she named All Ball, who was hit and killed by a car. She grieved for months.

March 10   A rainy night gave way to a clear day, and the welcome sun greeted me as I made my way through the zoo. It was still early, and I was alone, meeting nobody as I stopped every few feet to move soggy worms off the sidewalk to the safety of drier ground. Steam rose off the concrete surface as the morning sun began to climb into the sky. I thought about the strange human penchant for designating some animals as valuable and others disgusting. Kids would, I knew, jump on the worms to see their guts splatter while stopping in their tiny acts of destruction to pronounce the gorillas "totally cool" and to declare that they should be spared. I was amused by the passing thought that perhaps if someone were to write an epic on worms and call it *Annelids in the Mist*, worms might fare better.

As I walked down the slight incline to the shelter of the gorilla grotto, I knew something important was happening. The gorillas were standing stiff-legged and pursed-lipped, all staring at the same spot in the grass. Adhama slapped the ground and barked a threat in the direction of their gaze. I was starting to wonder if an animal from another part of the zoo had gotten loose and was hiding in the grass about ten yards from where the gorillas and I were standing. It went through my mind that I should run and get Karin in case it was a snake or some other thing that could be a threat to the gorillas' safety.

Then I saw a skinny, black, curled-up foot sticking awkwardly out of the grass. It was a dead crow. I leaned forward to get a better look at it. My movement toward the source of danger apparently upset the gorillas, who perhaps were worried that by some foul magic the crow might at any moment rise and stalk the living. Binami, Nonesha, and Taufiki began screaming as Adhama let out a deafening roar. I stayed absolutely still until they quieted and hugged each other.

They kept looking at me as if they wanted me to do something, but each time I tried to move, they became upset all over again. I decided to stand quite still and let events unfold without my input being interpreted as aviary necromancy.

After what I guessed to be about twenty minutes, Adhama began taking stiff, halting steps toward the crow. Between advances he would stop and pull up some grass to throw in the crow's direction. Occasionally, he would stop his forward advance to strut from side to side, his body completely tense, his eyes never leaving the bird's body.

After another twenty minutes, he had managed to advance to within five feet of the carcass. There he stopped, staring. Two minutes passed, then three. My leg was cramping. I barely dared to breathe. I knew that if I moved now, his fear would likely be turned on the other gorillas; he would charge them and slap at them, not really hurting them but upsetting them for the rest of the day. Of course, the day was probably already ruined for them. Gorillas do not recover from being upset very quickly. After five minutes of this breathless silence, Adhama did something completely unexpected: he buried the crow.

Digging handfuls of grass and dirt from where he stood, he

pitched them onto the bird with varying degrees of accuracy until even its gnarled foot, pointing accusingly, was no longer visible. The gorillas resumed their daily activity, though in a more cautious manner, studiously avoiding the fresh grave so carefully constructed.

"Comfortable hole bye," I thought, and so it was for the crow.

March 11   I received a message at the gate of the zoo entrance that Binami was quite ill from an unknown affliction. I was to check with Karin at the gorilla office for an update on her condition and to get some additional instructions in regard to my observations of the gorillas during her illness.

I had been to the gorilla office a couple of times before, to report the odd incident of visitors throwing things such as popcorn containers or toys into the habitat, and I made my way without thinking to the huge double doors screened from the public areas. I pushed the buzzer hidden in the wall and regarded the nearby Malayan sun bears eating grubs from a rotten log they tore open with their three-inch claws. I heard Karin's footsteps approaching. One thick door squeaked open, revealing her diminutive frame, made smaller by the contrast of the heavy door. She was smoking a cigarette, and her brow was creased with worry.

"What's wrong with Binami?" I asked right away.

"She won't eat. Not even my special concoctions. She drank part of a strawberry milkshake this morning, but that was it over the last 30 or so hours. I think she has some pain in her abdomen, and I want you to watch her closely today to try to get an idea exactly where the pain is, what's going on with her."

As we walked back to the office area she turned her familiar cigarette case over and over in her free hand, which was stained with fresh strawberries. As we entered the office, she put the case down on her desk and took a deep draw on her cigarette before crushing it out in a souvenir ashtray from Nairobi. She looked around at the

Binami

gorillas, who were peering into the office area from the grated windows of their individual night rooms.

"I have to take a stool sample down to the veterinary department, and I need you to stay here and watch the gorillas," she said matter-of-factly. I tried to seem nonchalant; I was thrilled.

She took out another cigarette and her lighter and went back out the door. She turned around as she got to the double doors.

"Feed Adhama some strawberries while you're in there," she called back to me. "But be careful not to put your fingers through the grating. He could rip one right off, and a bloody stump is never funny, you know! It's all fun and games until someone has a bloody stump."

I turned toward the gorillas as I heard the door squeak and then loudly clang shut. I was alone with the gorillas. The silence was deafening. The office was kept hot, and the smell of hay and gorillas was cloying. Gorillas smell like rhubarb pie, and combined with the alfalfa in their rooms, the tart sweetness palpably hung in the air.

We looked at each other. After breathing in their nearness for a moment, I took the strawberries from the sink and walked to Adhama's window. A four-by-four-foot grate of wire mesh was all that separated us as I stood self-consciously a few feet away from him. I stood quietly without moving, staring down at the cement floor. I couldn't look at him right away. It was not that I was afraid, I was . . . praying. Not the kind of prayer where you ask for something or try to find god. It was the kind of prayer you pray when you are born, or right after you are not in pain anymore. The kind of prayer you pray when you realize you are alive and surrounded by life. A prayer you pray for everything to be what it is. I was thinking that this was the kind of moment I would always live for.

Adhama was watching me with an even curiosity. I looked up to see him looking at me. He grunted gently, almost without making a sound. I took in his raw presence: five hundred pounds of muscled maleness. I began to see him as a whole being as the screen between us disappeared.

I feebly shook the bowl of strawberries, holding the bowl close to the grate, and I felt like falling backward as he raised up his bulk and pushed himself to his full height. He peered into the dish with the unself-conscious expression of someone looking at a dessert carousel to see if anything it holds seems interesting. He moved to the window and hoisted his imposing frame onto the stoop under the screen. He was now six inches from my face.

I began lining up strawberries on the window ledge so that he could reach them from his side. I became so absorbed in the task of balancing them in a row as quickly as he could eat them that I forgot to pay attention to where my fingers were in relation to his. As I placed the last strawberry on the shelf, our fingers touched as he reached for the ripe, red prize.

I stood agog at the sight of my tiny digit under his black sausage of a finger. It was rough and leathery like an old and loved shoe, and warm . . . tender. I quickly looked up into his pan-sized face and almost burst out laughing at his expression. His heavy brows were raised as high as they could go, and his eyes were bright with surprise. Simultaneously we looked down again at the ledge. For a timeless instant we were different sides of the same mirror: five million years of evolution joined at the finger.

We both jumped and moved away from the grate as we heard the telltale creak and clang of the door and Karin's steps drawing near. As she came through the office door I put the empty bowl in the sink.

I hoped she did not see that I was flushed. I made a pretense of rinsing the bowl as I let the cold water bring me back to the duties of the day. After I dried my hands, Karin handed me a stopwatch and a special check sheet for Binami and shuffled me out to my place on the oak bench. She would be happy to know that I still had all my fingers. But though I had not lost any blood, after my flush drained away there was none in my face for the rest of the day.

March 16 For the last several days Binami had been ill, and there was growing concern for her and the baby. As I watched her now lying in the hay of the grotto, it was clear she was in worse pain than yesterday. Every few seconds she bunched up her face and held her breath, sometimes for as long as thirty seconds. She was losing weight, down from 275 pounds to 250. I was hoping for a miraculous recovery. I was also hoping that it was not kidney disease. Renal failure and respiratory problems are the two leading causes of death in zoo gorillas.

Every time her face contorted in pain, I wished I could lay my hands on her and heal her like one of those slick preachers on television, or better yet, like a medicine person who knew the ways of healing from the inside out.

I remembered reading a story an American woman wrote about being taken on a walkabout, or spiritual journey through the outback on foot, by Australian aboriginals. At one point during their trek, an aboriginal man fell and broke his leg. The leg was set and left to heal without a cast or splint of any kind. When the foreign woman expressed skepticism about whether the leg would stay in place, an aboriginal elder patiently explained that the bone had been in the right place for thirty years, so why shouldn't it remember where it should be and stay there once it was returned? I hoped Binami's body likewise remembered how to be well.

I timed her sharp pains at intervals using the stopwatch that Karin had given me. I also recorded on a special sheet from the zoo veterinarian how long she held her breath and any unusual behaviors she engaged in. I noted that while she was lying on her back she would hold onto a stick and prop her feet up on it, an unusual example of tool use that seemed to help ease her pain.

She did not move all morning. Both Adhama and Taufiki were growing agitated. Taufiki began to put his hand to Binami's bottom and would then sniff his fingers, looking at her with a worried face. He alternately groomed her and pushed at her with all his strength. Adhama sat closer to her than was his custom and grunted at her every few moments. She would not move.

This routine went on for several hours until Karin appeared on the top of the habitat wall with the midday snack pail brimming with rare treats: peanut-butter sandwiches, frozen juice-sicles, mangos, yogurt-covered pretzels, and dried pineapple.

Adhama and Taufiki, probably famished by their morning vigil over Binami, began running to the wall, followed closely by Nonesha. Halfway to what must have seemed the Promised Land, they turned and realized Binami was still lying in the grass. Their ambivalence was immediately obvious as they weighed their concern for her against the lure of so coveted a feast. Adhama let out a bark of frustration in Binami's direction. Her eyes were closed and her face twisted.

Stiff-legged and with their lips pursed in consternation, Adhama and Taufiki strutted back to Binami's spot in the hay. Once inside the grotto, the two males began flinging heavy sticks at her immobile form.

This behavior is what researchers refer to as "ritual display."[14] Both individuals and groups as a whole, whether wild or captive, engage in these intimidating outbursts of physicality.

For captive groups, these displays consist of running with a stick or vegetation in hand and, standing up tall at the end of the charge, hurling their projectiles forcefully as a climax. On occasion, Adhama or Taufiki would rip up fresh vegetation for this purpose. Screaming or roaring would sometimes accompany these actions.

Mountain gorillas break off and hurl vegetation and run and charge in addition to sometimes roaring and screaming during displays. In the wild, gorillas engage in displays during different kinds of emotional stress. They also become upset and display in what is called "bluff charging," resorting to actual defensive attacks only when the group is in danger and there is no other alternative.

Even when you know you are in no real danger it is a frightening spectacle to observe. It jars the system and inspires fear.

Adhama and Taufiki's charges certainly jarred and frightened zoo-goers, who were unaccustomed to such noise and kinetic force. Visitors who had minutes ago been politely asking after Binami's progress toward wellness were now openly outraged at what seemed to all appearances the beastly treatment she was receiving from her family. I tried to explain that they were actually doing what they could to help her, but the sheer impact of their running, throw-

Taufiki's fierce display

ing, and screaming outweighed my feeble exhortations that they were really showing concern.

For some reason I suddenly remembered reading a study completed by a graduate student who recorded the amount of time zoo visitors spent watching the gorillas. The average visit was five seconds. The hurried judgments of this kind of meager investment toward understanding the gorillas' behavior revealed itself now: "See how mean those gorillas are, Johnny? They'd get you, too, if you were in there." "That lady said that gorilla laying down was sick. Those other two are trying to get rid of her because she is weak and

useless to the group now. Survival of the fittest is how nature works."
"Those gorillas are fighting for dominance. That one they are trying
to scare is the smallest male."

Again I tried to illuminate that they were attempting to get her
up to eat and that they were really concerned. Someone told me they
had a strange way of showing it and bustled the children off before
things got uglier. I heard parents telling their children that scientific types were just coldhearted, plain and simple. I thought if they
would only stay a minute I would be vindicated and they would see
that I really did not have a cold heart and that in fact I was pretty
swell. I winced. I looked after them feeling like a puppy that had
peed on the rug.

Unfortunately, the situation did become uglier. Even I was surprised when one of the male gorillas' projectiles, usually thrown
to miss, hit Binami smack in the head with a reverberating clunk.
It was the turn of those watching to wince as they held their collective breath. A three-inch gash on Binami's forehead began to
ooze blood.

Adhama and Taufiki stood riveted, heaving from their exer-
tions, waiting to see what would happen. Binami put a hand to her
brow. She brought it down to see that it was covered in sticky red.

With a pained grunt she rolled over and dizzily got to her feet.
She started to stagger toward the wall. Adhama and Taufiki strutted closely behind her, barking sharply when she slowed down or
turned to look at them. She sat in front of the wall and, looking
sheepish, began to eat the treats Karin threw down to her. Once she
gagged and some peanut-butter sandwich fell from her mouth, but
a final warning bark urged her to eat and swallow everything that
came her way. I continued to take notes on this extraordinary example of "tough medicine," though I knew that no one who knew
gorillas would be at all surprised by it.

Male gorillas in the wild will free group members' hands or
feet from poachers' wire snares by sliding their large canine teeth
between the tightening wire and the trapped wrist or ankle until the

victim is able to slide out of the trap.[15] This skill is crucial for medical reasons, as gorillas who break free with the snare still cutting tightly into their flesh will either become crippled or, more often, suffer the ravages of gangrene and eventually die of infection and pneumonia.

Silverbacks will slow the pace of the entire group as they travel and feed so that sick or dying members can keep up, sometimes even sleeping near these members until they die. Gorilla males will also lead their groups back from necessary feeding to support and reassure dying members who cannot go on.[16]

I had heard that in the past Adhama had tried in vain to protect his own family from veterinarians coming by to carry out routine checkups, and I had seen him try to protect a little girl on the other side of the glass who was being shoved by a group of boys, pounding repeatedly on the viewing glass until they left her alone.

It seemed that, like humans who made a career of tending the sick, gorillas could be pushy but effective. At the very least, Adhama and Taufiki started Binami on the road to recovery.

March 29   It was seeking the gorillas' care for the sick at heart that I arrived early at the zoo. I had intentionally gotten there well before the gorillas were let out. As I got to the habitat, Karin was just bringing out fresh hay and aspen branches to scatter in the gorillas' shelter.

I poked my head around the corner of the habitat door and smiled weakly at the comforting form of Karin in her torn black coat.

"You've heard, I see," she said.

"Yes," I said, my eyes avoiding hers. "How many?"

"Ten." She answered. She continued throwing branches.

"Did they catch them?" I asked.

A long pause.

"No."

War had broken out in Rwanda. Several mountain gorillas in the Parc de Volcans had been killed in the crossfire of people trying to hold back their attackers long enough to pass into Uganda. The mountain gorilla population was in the low 300s. They could not afford to lose any of their members.

I felt for the people running for their lives. I knew I would do the same thing they had done. My ambivalence sickened me. I wanted to choose a side and feel the healing fire of righteous indignation, of conviction. As it was I was sick at soul for the people, for the gorillas, for the future.

"I need to see Adhama," I said with a lump in my throat. I did not dare look at Karin for fear of tears.

"The door is open. He's in the middle pen, the one with the see-through door. I had him in there to weigh him this morning," she said flatly. It crossed my mind that she was probably ready to cry, too.

I made my way through the heavy doors of the hallway to the office area. As I went into the office, I could see that the gorillas were all sitting by the windows to their night rooms. They were silent. Not a sound came from any of them. Binami gently rocked back and forth, with a faraway look, seeming to watch something happening beyond the walls of her room. Nonesha sat beside her, holding her hand. Taufiki sat motionless with his head against the mesh of his window. Did they know what had happened?

I dropped to my hands and knees and slowly crawled over to where Adhama was sitting on the other side of the industrial-strength chain-link fence of the middle room, where the scales were set in the floor. As I settled in front of him I looked up into his huge warm face. He looked sad and resigned. My vision turned blurry, and then tears silently spilled out of my eyes and dropped onto the identification badge on my researcher's vest. Still we looked at each other.

Adhama moved forward and put his massive, hairy shoulder against the wire and motioned me to lay my head there. I let my head fall softly on the place where his shoulder had been offered and sobbed. I could feel the long hairs of his shoulder brushing the side of my face. I felt guilty. Why should this gorilla, taken from his mother, mistreated as a baby, living in a concrete jungle, care about my pain?

I realized it was his job. He was a good man. It was what he could do. It was his inner dignity to care. Nobody could take that away from him.

I had a similar realization about myself. It was not a sudden revelation or epiphany; it was more a slowly growing understanding. Though it was painful, though it would lead me to engage in what many of my colleagues would call "bad science," though it meant fighting a losing battle for the gorillas, though the observation had been made before, it finally was true to me that we are all in cages of one kind or another. The dignity of caring and bravely showing it is something nobody can take away. It is a waiting possibility in all of us. It lives in the fleeing people, the crying people, the hairy people we call gorillas.

Ultimately, we know there is no one to blame. We are responsible for doing what we were made to do. Suddenly, Adhama and I were equals. I would never call him a victim again. . . . He was a silverback.

He grunted his reassurance, and through the minutes that passed, I felt my tears running away from my eyes and back into my body. It wasn't that I did not want to cry; it was more like my tears had become refugees, running for their lives over faraway mountains where gorillas had always lived.

It was spring, I thought. I'll leave the crying to the sky.

April 2 Binami was starting to recover, and I saw in the daily reports that it was probably due to the intervention of Adhama and Taufiki that her life had been spared. By forcing her to eat the treats laced with antibiotics, they had quelled the infection. She was lucky to have these men in her life.

I thought back on my morning of crying on Adhama's shoulder—feeling weak and strong at the same time—and pondered my convictions about feminism as I had embraced it during the 1970s and 1980s. I had certainly needed Adhama's style of strength that day. His sheer mass was comforting to me. As a woman getting a Ph.D., I was confident that I was capable, smart, and independent. It occurred to me that nothing of Adhama's comfort belittled or subsumed me. Clearly Binami's capitulation to the wishes of the males in her family had saved her life, though at other times before she had felt completely at ease trying to get the best food items before they did and had even ignored their wishes at times. It seemed that in a crisis, she and I trusted that they would look out for us and ask little in return. Thinking along these lines was counter to the feminist movement that was responsible for the freedoms I now enjoyed. The idea that there was a biological basis for these tendencies was suspect indeed.

I remembered hearing the story, brought out of the closet and dusted off whenever the fighting undergraduate women around me needed inspiration, that during the neolithic period in Europe there had been a golden age of matriarchy during which women were in charge, made all the major decisions, and even fought wars as fierce warriors who inspired dread in their male adversaries. As I began actually studying the neolithic, or "new stone age," two problems with this narrative surfaced immediately.

First, there was no such golden age. Evidence of an egalitarian society, wherein women and men were both respected and included on all levels of life, exists for the time and place in question; however, little or no evidence indicates that women took on "male roles" in the culture of ancient Europe or anywhere else on the planet in recorded history.

This brings us to the second problem. The golden age of matriarchy that the women were describing was not very creative: the women in the myths they were calling upon for guidance were basically taking on the same kinds of power and dominance (in a very Western way, it might be added) as the men of the modern patriarchy they were so adamantly opposed to.

It may be that they felt they needed to structure the myth in this way to be inspired to fight a system that was clearly in need of radical change, or that they felt as if they were getting some sense of power and agency back by aligning themselves with uncompromising archetypes of their foremothers. Whatever the reasons, the subtle point of these myths was never examined—the point being that someone needs to lead, to fight, to prompt, to protect.

I found out when I began to talk to women of color outside the university that white, middle-class, academic feminism was something they found hard to relate to. They had a deeper realization that balance is what we had all come to lose. They seemed to understand that strong men were needed to ensure strong women and vice-versa.

Many times I had heard rude, socially insensitive, or sexually aggressive men called "big apes" by both women and men alike. I had a growing feeling that "big apes" were great role models for men: not role stereotypes, but role models.

Exemplary male behavior abounds in the stories of gorillas' lives. One hunter was quoted by George Schaller as having witnessed silverback males assisting disabled group members to safety during dangerous encounters. Many tales recount heroic feats of ingenuity in which silverbacks extract individuals from snares and traps. Male gorillas in the wild routinely give their lives standing off poachers so that their families can escape.

Male gorillas are ever vigilant for signs of danger, and group members are constantly tuned in to the moods, demeanor, and body language of their silverback leaders; the silverback sets the emotional tone for the entire group as a result of his position as protector and leader. Males protect their groups from intruding gorillas from outside the group as well and drive other males away with bluff charges and actual attacks. Silverbacks post themselves as sentries at the group's edge during times of rest or sleep, and in fact they stay awake and alert, intermittently vocalizing and chest-beating deep into the night.[17] Silverbacks also settle squabbles and disputes within the group.

In an outstanding example of paternal behavior, they have been known to adopt orphaned gorilla babies, allowing them to travel by their side and even sleep with them in their night nests: the small and frail against the warm massive body of the protector.[18] In the wild, Schaller reported that while keeping vigil next to a dying companion, a single male gorilla defended his ill charge by attacking a threatening leopard, eventually driving it away and thereby prolonging the life his companion, if only for a day.[19]

As I sat watching the gorillas quietly forage tidbits from the habitat and glean contentment from each other on this quiet afternoon, I thought back on the odd, busy days of my research that had noisily passed as I tried to keep my thoughts clear and take accurate notes. I was suddenly sad remembering Adhama and Taufiki threatening the noisy crowds, throwing sticks at the public window, and charging with grimacing faces toward the unrelenting stream of maddening humanity, unable to protect their family and restore them to peace in mind and spirit. These impotent attempts at manly intervention would invariably bring shrill shrieks, nervous smiles, and barks of laughter from the crowds. Each of these behaviors was perceived as threatening by the gorillas and, as a result, caused more displays from them in an endless and deafening vicious cycle. The audible sound matched the roar of soulful discord, a jarring clang from within and without.

The Mbuti described dissonance of this kind as being a thing more damaging to the spirit than to the ears and felt that *Noise* was a thing both brought about and suffered by civilization. I think that it is this insidious noise that men of our species rage against to no avail and that women fear and wish that men could make go away.

If it were a mere matter of a tangible threat being confronted by tangible bodies, men would have scared the dark, yowling thing away long ago. For essentially all of what is known of our 65-million-year history as evolving primates, men have been bigger and more muscular than women, who have had in turn greater stamina and tolerance for pain and discomfort than males.[20] Males have stood and confronted, sometimes at the cost of their lives, while women have scooped up their young and have run until they collapsed. Confrontations have been decided with decisive and deadly swiftness, while a woman's stamina has left a long, evaporating trail.

Both kinds of strength have been absolutely necessary. If both had identical skills of perception, reaction, and physical prowess, all would be lost. If both were not sufficiently similar as to see the threat, however, all would be lost as well. As men and women today we see the threat that is poised to descend on us. The problem is that we see it reflected in each others' wide and terrified eyes and are unable to do our jobs with the seamless precision that the balance of male and female, confrontation and escape used to secure for us.

April 7   The sun was high and bright in the sky as I arrived in the grotto to see the gorillas beginning their daily amble to the wall for afternoon snacks. I was relieved to see Binami back in her place in the slowly moving group, her eyes bright and expectant as she kept Karin and the metal bucket fixed with attention. Something out of place, though, was the burlap bag around her shoulders.

During Binami's illness, she had torn apart the seams of a burlap bag left in the habitat with treats inside. Stretching it out into a long rectangle, somewhat like a scarf, she had wrapped it snugly around her shoulders. We all guessed that it was helping her stay warm.

This unprecedented creative use of material was quite exciting to me as a researcher interested in the development of culture.

Though there were obviously never any large dead animals or animal skins in the habitat for the gorillas to drag or carry, this use of burlap had not been seen before in captive gorillas; it gave insight as to how our early ancestors might have begun using materials available to them in these ways.

Mountain gorillas have been seen to drape cushions of moss around their necks and walk around with them, apparently enjoying the activity. Given their often cold, misty habitat, perhaps they too enjoy the feeling of an improvised muffler, carrying it around on their necks for long periods of time and replacing it in its proper position when it slipped.[21] Dian Fossey observed gorillas carrying food, including whole branches of fruit. She also recorded young gorillas harvesting bracket fungus and transporting it several hundred feet from its source high up in the trees.[22] Not as uplifting is Fossey's description of the dragging of a dead gorilla's carcass by its distraught family members.[23]

Carrying things around was certainly not in itself unusual. The gorillas in this study exhibited a wide range of carrying behaviors. Many times they carried tools, including sticks for display, hunting, reaching, and digging, as projectiles, and as rain parasols. Binami still carried Nonesha occasionally, and the gorillas were seen to carry nesting materials by the armful to distant locations in the shelter. As I had seen many times in response to noisy crowds, gorillas carry things to throw, too. Male gorillas in the wild uproot trees or break off branches to use them as display enhancers or projectiles, transporting them from one location to another before releasing or throwing them.

I could not picture Binami throwing away her now-beloved blanket at the end of a charge, no matter how angry or frightened she was. As I watched, she paused in her trek to the wall to arrange it daintily so it hung evenly over her shoulders, which were still a little withered from the weight loss she had suffered over the past weeks. She continued toward the wall, still breaking her stride every few steps to fiddle self-consciously with the light-brown cloth. As

Binami with her beloved burlap blanket

she arrived at her usual place in front of the wall she circled around her spot and began picking up dead leaves and other debris blown into the habitat. I was wondering if she was planning to eat them as the withered vegetation accumulated in her hands. I mused that she would have to be hungry indeed to consider these bits appealing. Certainly she had some other goal in mind. I really began to wonder, though, as this went on for a time without sign of ceasing and with no fragment missed. With a swift motion she hurled the debris in her hand downwind and seemed to watch with satisfaction as the breeze caught the pieces and took them away.

With a grunt she turned her gaze to inspect the area around her for any trash that had escaped her scrutiny and looked assured that the spot was free of refuse. She stood up and with a flourish snapped the burlap scarf from her neck and then let it billow in front of her, holding two of its corners in a pinch of thumb and finger. The material shed the few pieces of hay that had been clinging to it. Binami then lowered the burlap, still billowing, down to the ground. It lay almost perfectly flat on the grass. Binami scooted around its perimeter to smooth out wrinkles and flip a corner that had gotten caught in the wind and remained flung back toward the center of the rectangle. When the material was perfectly spread like a fine picnic blanket, Binami plopped herself down in the middle of it and looked up to Karin after letting out a long sigh.

She seemed suddenly self-conscious as she noticed the look on Karin's face and then looked around her to see the gorillas evincing the same expression. It had been a shockingly involved production, and we were all stricken and, I am sure, all looking a little stupid from Binami's point of view. Determined to get lunch rolling and take our stupefied attention off herself, Binami fixed Karin with a look of riveted interest and acted as though nothing had happened.

With her mouth open Karin looked over to where I was sitting on the other side of the yard. I shrugged and gave her a twisted smile. I was becoming less and less amazed by anything I saw. Karin shrugged back with a delighted smile and began throwing celery stalks down to the gorillas—much to Binami's relief, I think.

April 15 I arrived at the grotto at about the same time as yesterday, though the sky was full of threatening clouds. I sat on the wooden bench and rummaged quickly for my pen and clipboard, deciding to pour a hot cup of broth from my Thermos right away as a safeguard against the unseasonable cold. With the clipboard on my chilled knees I peered through the steam rising from the cup of broth nestled between my hands.

Karin must have started their lunch ahead of schedule, as the gorillas were almost done, still in their customary positions in a semicircle around the base of the high wall. Binami had apparently adopted the burlap picnic blanket, if not permanently, at least for the time being. Karin, taking this new preference in stride, was smiling her usual warm smile and talking to the gorillas using the same familiar and sometimes nonsensical words and phrases she had uttered to them since they were babies.

As lunch came to a close, Karin threw down the last item on the menu, which happened to be shredded-wheat biscuits still in the long envelopes of paper they were packaged in. Normally, the gorillas eagerly tore these packages apart and the envelopes were immediately discarded in favor of the contents, a favorite with them all.

All the gorillas but Nonesha tossed the paper aside, true to form. Nonesha, however, seemed today to be more interested in the paper. With a look of impenetrable concentration, she painstakingly tore the paper along the seams and, laying it on the ground, smoothed it out with her palms until it was reasonably flat. She sat back and looked at it critically. With pursed lips expressive of a frustration bordering on histrionic, she pressed the rectangle until it was even flatter and then sat squarely on the middle of it. Only at this point did I realize that she was imitating her mother.

Even before this singular instance, the gorillas had readily demonstrated capacities for imitation, learning, and creativity. Many examples of learned behavior through imitation had, in fact, involved Nonesha learning social and life skills from Binami. For instance, Nonesha had attempted to build a nest at the same time Binami did. Once she mimicked a charging display immediately after Binami had displayed, the fact that her little frame made her fury more comical than convincing notwithstanding.

Examples of meeting new challenges creatively had been illuminated in such instances as Nonesha's employing a combination of tools to hunt and using a stick to hammer raisins out of a "raisin

Adhama taking a break while prying raisins from a log

log." All the gorillas had learned to use sticks as poles to obtain food or other objects out of reach.

When the logs were first introduced on November 16, 1991, for instance, Adhama was the first to discover he could get them out by using a small stick. He first used a stick that was too big, then tried a thick spear of zucchini before finding a small stick of the correct size.

Seeing Nonesha's clear and unusual copying of her mother's behavior, even when it was idiosyncratic, opened a window onto how such purely aesthetic preferences gain a foothold in a culture. By this point in my life, I had heard many people protest the notion that we as humans evolved from the apes, even though the theory of evolution never posited that we had evolved from gorillas, but that we had a common ancestor. As I watched Nonesha sitting with

Binami using a small stick as a tool

supreme confidence on her tiny square of shredded-wheat paper, I had to wonder if our common ancestor had a deep aesthetic bent. I knew as I was thinking about it that people who had trouble with our shared history with the apes would certainly never believe that our common ancestors might have appreciated the sensual appeal of a picnic blanket.

May 24 Play, even in its primal manifestations, could be considered aesthetic in and of itself. I watched Nonesha and Taufiki out in the habitat. It was becoming rare to see them play with such abandon, for they were getting older and Taufiki had even begun to take over some of the dominant male responsibilities that until this year had been the exclusive domain of his father.

Seeing him laugh his panting gorilla laugh as he reached toward Nonesha with exaggerated tickling motions, it was hard to believe that in the wild he would probably have left his group by now in search of females with which to start a family. Gorilla males, in their natural habitats, will break from their groups around puberty. Occasionally young males will stay in the groups of their birth to form an alliance with aging silverbacks, making formidable teams, but most emigrate out of their natal groups and remain solitary for a time or join all-male bachelor groups comprising other young males.[24] Individual males eventually strike out and begin the lifetime process of persuading females to join their groups.

There are several reasons a female might decide to take them up on their offers. A female of lower rank in a well-established group might decide to try to improve her situation by being among the first to join a male's new group. Generally, rank among females reflects the length of their association with the silverback male, or the order of their pledged membership.[25]

Groups that have lost their silverbacks through death or illness are lost until they find a new leader. Large groups, however, are hard for young and inexperienced males to hold together, because it takes a high degree of prowess and wisdom to lead well. Sometimes these groups split up and offer the new males opportunities to expand their groups, although many times a male next in line of dominance in a group will step up to take responsibility.

Once a male has females in his group, he still must fight to keep them; it can work for or against the young males if a female decides she finds a silverback outside her group more attractive. All in all it is a tough process for young males, though it is certain they accept it as a fact of life and even enjoy the challenge. They need to go forward and want to assert their agency—make their contribution.

Taufiki was not going anywhere, though. Not now and perhaps not ever. Since the institution of successful breeding programs, there has been a shortage of females and a surplus of males among the captive gorilla population. No one had expressed any interest in Taufiki, and that seemed unlikely to change. Sometimes he appeared to have some kind of awareness of his grim position, for he spent whole days upset and displaying at his family and at the crowds, even when they were sparse and quiet.

This was definitely not one of those days. He was in a positively giddy mood. Comically, he tried several times to hide his massive 350-pound frame behind puny little stalks of nettle toward the back of the habitat. In good sport, Nonesha pretended not to see him and feigned immersion in searching for bits of food in the grass. She would sneak glances at her brother as she moved forward, her search unerringly taking her within his striking range. As she drew nearer his huge muscles would bunch in anticipation just before he pounced on her with a loud roar. Getting her down on the ground and straddling her, he would then mercilessly dig his thick, short fingers into the sensitive areas of her neck and underarms until she was breathless from being tickled. With her wide mouth thrown open and her eyes squinted tightly, no sound would come out of her mouth until she would draw in a frantic breath and buck Taufiki off her big barrel of a middle. After that she would break free and run away, settling beside Adhama when she reached the shelter. Adhama had been watching these capers and would play along by putting a reassuring arm around Nonesha's shoulders until they saw that Taufiki had once again taken his not-so-clever perch behind the nettles, whereupon the whole cycle would start over.

Eventually, Nonesha tired of the repetitious game and ignored the grinning Taufiki as he trembled in anticipation behind his nettle stalks, which were now in such a sorry state from ambush after ambush that only one of the three stalks was left in anything resembling an upright position, and even it was bent at right angles toward its middle. Maybe she felt that she really couldn't go on with the charade at this point in the game.

Nonesha decided to reverse the game. Casually she strode to the large oak tree right outside the shelter and, slipping behind it, peeked around the trunk just long enough to mark Taufiki's post behind the nettles before she pressed her face tightly against the bark of the trunk away from his view. She waited in this manner for several moments. She seemed to have forgotten that keeping an eye on your approaching quarry was an integral part of the game, for when she peeked back around the trunk the object of her plan had disappeared.

Adhama and I, sitting close together in the corner of the shelter, were both stifling laughter. We had seen that Taufiki had taken Nonesha's carelessness as an opportunity to circle around her. Her look of perplexity turned to shock when she moved to resume her hiding place and Taufiki, seemingly out of nowhere, roared with all his might and grabbed for her. Seeing a completely shocked gorilla weighing 175 pounds leave the ground like a rocket is truly something to behold. At the apex of her trajectory her arms gave a small convulsive shake and she made a strange strangled noise as her tongue stuck out stiffly from her grimacing mouth. Her eyes bulged out. By the time she landed, stark terror had turned to fury, and like never before or since, she gave Taufiki the beating of his life. As she bit his arms and pummeled him with her fists, barking angry barks, Taufiki kept turning from the fray toward Adhama and me with a puzzled expression as if to say *What? Wasn't that funny?*

Creative, interactive play was as frequent in Adhama's family as it is in wild populations. Binami sometimes imitated the comical gestures of zoo visitors. Adhama had been seen to use a stick to draw in the dirt. And, like today, Nonesha was often engaging Taufiki in spontaneous, evolving games like "hide and seek." Nonesha also regularly played with objects as toys: hay, burlap, browse, and sticks were all more to her than simply utilitarian.

Specific examples of the enthusiasm for creativity and imitation during play in wild communities include a young gorilla who invented a display using a branch with fruit on it and another young gorilla who learned to look through a researcher's binoculars (through the wrong end, as those lenses better fit his wide-set eyes). He would wiggle his seemingly distant fingers in front of the lenses, dropping the binoculars quickly away to verify that his fingers were still attached.

Nonesha may have wished that she had been observing Taufiki through the wrong end of binoculars today. It might have saved wear and tear on her heart.

May 25   After yesterday, it seemed clear that Nonesha was not in the mood to play. Perhaps it was only her pride that was wounded, but she stayed close to Binami all morning. As unusual as Taufiki's jubilant abandon was yesterday, Nonesha's clinging to Binami was just as out of character since she had begun to mature this year. Not long ago, however, Nonesha's habitual nearness to Binami was to be expected. Binami and Nonesha had exhibited close proximity nearly all of the time before last fall, even though Nonesha was around five years old at the time. Though she would sometimes move away from Binami to play or explore, she frequently came back. Any time she was frightened, for example, she returned to Binami's side.

Binami taught both of her offspring social rules and living skills, more intensely to Nonesha. Reprimands for breaking these subtle rules included warning barks, palpable ignoring, and, as a last resort, quick bites meant to shock rather than hurt. Bites were used as a first line of defense, however, when Nonesha would not stop playing on Binami's stomach.

Mountain gorilla mothers show similar patterns; they are nurturing and solicitous starting from the births of their offspring but even with two-month-olds will enforce gentle discipline by mock-biting or repetitive pushing.[26]

Many times when Binami doled out discipline, it sprang from worry more than annoyance. Twice Binami angrily drove Nonesha away from the habitat pond, clearly feeling that Nonesha had gotten too close; her behavior was evidence of Binami's awareness that because they are nearly all heavy muscle, gorillas sink in deep water. Habitat designers had found out the hard way that gorillas are not anatomically suited to dealing with the dangers of water. Many zoo gorillas have been lost to moats designed to keep them safe inside their habitats.

In keeping with these protective impulses, gorilla mothers are as keen on having their offspring near as the offspring are keen on clinging. Binami routinely carried Nonesha ventrally (on her stomach) when Nonesha was a small baby. They traveled together throughout the habitat during episodes of male herding when Adhama led them around the perimeter of the habitat.

Offspring in the wild also ride ventrally for about a year, after which they ride on their mother's back most of the time.[27] As in Binami and Nonesha's case, the importance of proximity is demonstrated in the offspring's reluctance to travel independently. In both captive and wild gorillas, this results in rump-clinging behavior, which, as a final stage of mother-assisted travel, consists of the youngster's clinging to the mother's rump hair while walking on two feet behind her.

Enacting another common behavior in gorilla mothers, Binami often took food from her small daughter. This could have been an attempt to teach her some dietary rules, a way of demonstrating her social dominance and gaining choice foods for herself, or a combination of both. It was noted that Nonesha did not protest when her mother took her food but wailed when Taufiki tried to take things from her. Sometimes Binami would insist that Nonesha surrender whatever it was that Taufiki wanted. Because this horrific trauma (from Nonesha's point of view) usually occurred after the mother and daughter had retired to Binami's nest with a store of goodies, Nonesha (when she saw her behavior was getting her nowhere) would pitch herself tragically on the rim of the nest, her face pressed deep into the hay.

Young mountain gorillas continue to night-nest with their mothers up to five years of age.[28] This intensity of contact is observable throughout the offspring's development. Almost all infants maintain constant contact until five months and still at thirty-six months spend 30 percent of their time on their mothers. Strong social bonds persist between mothers and offspring long after mothers have ceased to provide food and transport. Dian Fossey has detailed the ways in which mothers teach offspring social rules and keep offspring from annoying other individuals, even noting that mothers seem embarrassed when children overstep social boundaries.

Wild mothers also intervene to take dung and other nonfood items away from offspring and inappropriate foliage away from infants. As young gorillas become older, however, their mothers are more lenient than many human mothers when it comes to letting their children explore and experiment. Maybe this is because adult gorillas in general are quite sedentary and mothers may feel that the expenditure of energy is best saved for real emergencies rather than constant small-scale intervention. Or it may be that they find their children's experimentation intriguing and engaging and, like the best of human parents, feel a vicarious joy in the exuberance of youth.

May 27 The day was cold and rainy, and I was sure as I made my way through the paths to the gorilla grotto that I would not see the gorillas, even the younger ones, interested in playing or experimenting. I imagined they would huddle inside the grotto, casting chagrined looks out to the yard where everything was silently and steadily dripping.

I was not too happy about this day myself. The weather had been gloriously warm and sunny, and I thought we might have the rare luxury of a relatively dry spring season. I put a folded blanket down on the damp observation bench. I had come to think of this bench as a second home. Its fragrance surrounded me; elements of old wood and moss came from its heart on these wet days, making it smell like an old cabin. Its scent mingled with the rich woodsy, floral, and barnyard smells of the zoo around the grotto.

I took out my clipboard and set it beside me. On it I put my Thermos of Earl Grey tea while I began to pick at the steaming raisin bran muffin I had bought at the concession stand on my way down to the shelter. It was going to be a slow day. I felt smug in the conviction that I knew exactly what the gorillas would do on any given day according to the weather.

My old friends began coming out into the habitat. Adhama stopped in the doorway to the yard and looked up to the sky with his eyes squinted against the drops that splashed off the metal sill into his upturned face. He looked down at the soggy, muddy mess outside the door and put on a disgruntled expression. He gingerly put out one great hand and then a great foot, curling his lip as the brown pudding that was the trail oozed up between his fingers and toes. Like a swimmer inching into a cold lake he slowly brought his other foot and hand out and eased them into the mire. His lips were now pursed, and he seemed to struggle with letting go of the warm, dry night rooms. Resolved, he stiffly strutted toward the shelter, shaking his hands and feet every few steps.

Taufiki was right behind him but had no thoughts of gritting his teeth for the inevitable. His habit on wet or snowy days, beginning as soon as he could walk, was to grab the sides of the door to the yard, give several back and forth tugs, like an Olympic skier preparing to exit the starting gate, and explode toward the shelter, running upright the entire way. Once inside he would always grab a handful of dry hay and wipe himself off quickly, and this is what he did today.

Binami was far more stoic than either of the men in her family. Without hesitation she plodded out the door and to the shelter, normally with Nonesha hanging onto her rump and walking upright so she could keep her hands clean, much as her brother did. Today

Binami made her way to the shelter alone, and I wondered where Nonesha had gotten left behind.

Without warning Nonesha came bounding out of the door with an enormous grin. She tore across the back of the yard and disappeared behind some of the trees in the back. This was new. I scrambled for my pad, knocking over the Thermos and losing my muffin in a shower of crumbs as it bounced off my knee and rolled into a puddle. I started scratching out notes and times without taking my eyes off the trees at the back of the habitat, which were now shaking violently. Nonesha had climbed a young birch tree and was forcing the whole top of the tree to wave widely back and forth as she shook it with all her might. As the tree swung wider and wider with the building momentum, I became concerned that the tree would snap and Nonesha would come hurling down to the hard ground.

She was making a whooping sound unlike any vocalization I had heard. The whoops sounded like a wild drunk woman out of control at a party, the kind of exuberant laughing yelps such a partier has just before she looks funny and tries to find the bathroom. I was worried that Nonesha might soon meet a similarly unpleasant experience if she did not stop swinging from the tree. I grimaced as the tree bent nearer the ground with each wild career.

Finally, my fears were realized as a crack that I thought at first was lightning turned out to be the top of the tree coming off with Nonesha still attached. She sailed in an arc from fifteen feet up in the air to the ground below, seeming to hang suspended in midair for several seconds, her legs flung far apart.

I saw the underbrush receive her sailing form, and then all was still. Only a few leaves jarred from the tree lilted down with the rain and bore witness to the frenetic activity that had just taken place. She's dead, I thought.

The gorillas looked at me and looked at each other, and then we all looked back to the underbrush. I believe they were having similar thoughts regarding Nonesha's untimely end. Then, just as suddenly as she had exploded like a cannon from the door, she exploded from the underbrush, still holding the top of the birch tree and dragging it behind her. She headed straight for the shelter.

With a wide-eyed attitude that I had never seen before, even Adhama and Taufiki cleared out. Prized celery and baked apples were unceremoniously abandoned as Nonesha came bursting into the shelter, dragging and waving ten feet of tree in her wake. She careened around the perimeter of the shelter at full speed and then headed back out into the rain. With a deep grunt that turned into a roar she heaved the treetop into the air, and the other gorillas, who had hurried back into the shelter, poised to run away again as the leafy monster looked as though it might land back in their midst. They stood heaving and bewildered as the projectile came to rest just outside and Nonesha leaped into the shallow stream running through the habitat. Waving her long arms to and fro, she sent waves of water splashing in all directions with her mouth wide open to catch the sprays coming down from above. She straddled a log that joined the two sides of the stream and poured water on it with her cupped hands. After doing this several times, she beat and rubbed the log with both hands, her whooping keeping time.

Then, as if someone had pushed a button she went limp, lying face-down on the log, her arms and legs dangling limply on either side of it. She remained like this for a full minute. She then dismounted the log, careful to avoid the stream, and sauntered back to the shelter. It was as if nothing had happened.

Jane Goodall, having watched similar "rain dances" among the chimpanzees of the Gombe Reserve in Tanzania, speculates that these wild displays are very old and find their beginnings in an ancestor the apes share with humans: displays that gave rise to the elaborate rituals and ceremonies we still associate with the changing of seasons and the marking of time, of patterns in life.[29]

R. Jane Eisler concludes from supporting evidence, especially that of primatologist S. Kuroda, that early hominids indeed possessed the unified social behavior that engendered ritual displays.[30]

Intensive studies on the types and context of basic vocalizations such as roaring hypothesize that ritual displays accompanied by vocalizations were present in common ancestors of apes and humans.[31] The climatic seasonality that the Miocene hominids would have been witness to could have inspired such displays in the face of dramatic elemental occurrences such as rain, thunder, and lightning. Theologian Joseph Campbell concedes to an early genesis for ritual, perhaps previous to the ape-human split in the earlier portion of the Miocene, citing Wolfgang Kohler's conviction that ritual involving elaborate motion patterns is a very ancient behavior, perhaps marking not only tangible events but rites of passage as well.[32]

As Nonesha peeled herself from the log and approached her mother, resting her head on her mother's shoulder and her hand on Binami's huge pregnant belly, I wondered if she, too, had realized that a time in her life was ending and another was about to begin.

July 2 Yesterday morning, at 2:00 A.M., Malaika came into the world through Binami in the same corner in which her siblings had been born. The heat had been turned up in the office and night rooms, and the usual bright lights had been replaced by red ones, giving the humid rooms a soft primal and comforting glow. Karin and I had been joined by several zoo staffers and the veterinarian when it was clear that Binami would soon have her baby. No one had spoken for several hours, giving the event an enfolding and religious aura—not oppressive or stifling but gentle and sacred. We all shared the heat, the dark, and our breath.

The birth had been a fast and easy one for Binami. After three hours of labor, during most of which she was still interested in eating, she pushed a sumptuous mound of sweet timothy hay into her corner and sat on top of it.

She leaned back into the familiar support of the corner and let her legs follow the angles of the walls. Her eyes were closed but

Malaika

only pinched tight once as a gush of liquid issued forth, followed by the tiny top of a head with its wispy hair plastered down with wetness. Another push and Malaika squirted out into Binami's waiting hands. Immediately Binami pressed the tiny, shaking baby to her breast and began to lick the top of her head.

Malaika blinked widely at her mother and tried to hold her mother's face in view for a few moments, then began rooting around for a nipple. Binami let her little daughter slide down and attach to her breast, and Malaika began to nurse contentedly, her minute pink hands pressed to the dark hair of Binami's chest. A cheer went up from all of us who looked on. This would be a healthy baby.

Adhama, who had remained at his usual post in the doorway,

had watched the process with an expression of wonder, and when we cheered he looked up at us brightly and grunted his satisfaction along with us. He did not attempt to approach the mother-and-child pair in a show of support, as that is not the way gorilla fathers do things, but he rested his big, shaggy head in his hands as he was propped up on his elbows and peeped through the doorway, completely transfixed by his new daughter. Binami grunted back at him. Nonesha and Taufiki pressed their faces against the clear door that separated them from Binami's room. They were riveted to the squirming, mewing figure in their mother's lap—completely pink and almost bald—occasionally trying different angles to get a better view. Their hot breaths made pulsing traces of moisture on the window.

Not a soul moved as we watched the family and the gorilla nation welcome the five-pound addition into their lives. In my heart I welcomed her to my life and my nation as well and promised as a birthday present that in some way my work would help to ensure her a better life than many of the people of both our nations had endured.

Adhama lifted his gaze from his daughter and looked into my eyes as if he had understood my thoughts. "I promise that to you, too, my old friend," I whispered, and we both looked back at the tiniest burden we had ever loved together.

July 5   The birth could not have happened at a more fortuitous time, from the perspective of the zoo administration and the zoo-going public. The heavens seemed to have taken a hand in what could be considered the summer season's first theatrical production—Malaika's grand entrance into life behind the window—and the sun cast a bright, warm spotlight on the stage of the attraction.

Though it was early in the morning, crowds pressed into me and jostled for the best views even though the gorillas had not come out yet. One man stepped in front of my bench, though the space in front of me was too small for anyone to fit. He lost his balance trying to make the squeeze and fell over on me, sending my notepad flying and crushing my foot while trying to right himself by pushing on my shoulders.

I had bent down to retrieve my notes and pen when I heard a buzz go through the assembly. Adhama had appeared over the hill. When he saw the crowds he turned and tried to get his family back into the night rooms. He seemed to be understandably concerned for Binami and Malaika.

Binami had been held inside for several days, however, to make absolutely sure that all was right with her and her offspring. She was more than ready to get out into the air and back into her familiar routine. She pushed past him without courtesy as he tried to block the doorway. When Adhama realized that his herding efforts were futile, he capitulated and led the group to the shelter and the expectant throng.

Though Binami moved eagerly around the habitat and held the baby close so that it was hard for anyone to see anything, after she had made the rounds to see that the habitat was in order she finally sat down near the shelter window. The little gorilla started to squirm and protest. After an eternity, during which the visitors collectively held their breath, a pink face the size of an orange peered over Binami's long, hairy arm. The newborn blinked her pale eyes sleepily and then yawned a tremendous toothless yawn with her Lilliputian tongue sticking out. She smacked her lips and then gazed into her mother's tender face before burrowing in to find a nipple once more. Amid squeals of delight and exclamations proclaiming her'a darling, it was clear that the public, at least for now, was in love with her.

Malaika, now a separate and adored entity, got her name after the zoo publicity department cooked up a contest; the person responsible for submitting the name finally chosen by the zoo staff would get a cash prize as well as a behind-the-scenes tour, culminating in a five-minute visit with the baby. This process disturbed me for a number of reasons. First, the cash prize I felt was tacky, but that was an aesthetic hang-up on my part and really of no consequence. Second, and more reasonable, was the fact that zoo workers had to be regularly tested for harmful or potentially fatal diseases that could be passed to the animals; the winner of the contest would not have to undergo these precautions before coming face-to-face with the baby.

But it was the third point that really bothered me. It was not about health issues or tacky pageantry. It was the fact that zoos never named the babies who died. The contest winners were never announced until it was determined that the baby would most certainly survive. I realized that winning a contest under the unhappy circumstances of a baby gorilla's dying in infancy would be less than something to celebrate, but somehow it seemed wrong not to acknowledge that they had lived; people had anticipated their arrival, hoped for their future. Mothers like Binami had carried all of their babies, whether the infants survived into adulthood or not, and I am sure they felt their losses keenly. In the end, though, lost babies carried only numbers.

As I watched the wriggling infant at the center of our attention, I thought that numbers, by which we divide the world, had a lot to do with our attitudes about life and death. Since humans developed the ability to delineate the world we have indulged in a process of dividing and conquering. This is a skill that obviously has paid off in terms of evolutionary success. I would not be able to be a scientist or exist at all if my ancestors had not picked up this talent and run with it. But trailing behind this talent for numeration as we ride it into space, paradoxes continue to remind us that categorization does not come so easily.

As we categorize things—living/dead, person/nonperson, good/bad—we must make uncomfortable nods toward context. Evolution. The idea of *eventually*. No one would ever question that 1 + 1 = 2, for instance, but when you start asking, "One *what* and one *what* equal two of *what*?" you get into deep mathematical waters right away. For instance, one cow plus one bale of hay eventually make a gallon of milk, or a healthy calf, or enough sperm to eventually make a million cows, or perhaps you end up with one very fat cow. One cow plus one bale of hay also could represent the eventual product of billions of years of evolution wherein bovines and grasses have influenced each other in a completely integrated way, and in that case 1 + 1 is already the answer; there is no 2 to be added.

In the process of evolution, billions do, in a very real sense, equal 1. You. Or me. Or a gorilla. But at the same time, and regardless of how long we live, we need everything weaving around us as the weaving needs us in return.

We are not the only living things on the planet that struggle with the chasm in our knowledge that the world can be divided up and yet, somehow, cannot. Researchers have recorded whale songs that were identical in every way but one: each whale ended his or her song in a distinct way that remained the same each time. The researchers were at a loss to account for this minuscule but persistent idiosyncrasy until they realized that the signature notes at the end of the song were the whales' names; the whales understand that they are separate from each other. Yet they are not so separate that they can live apart from the sea. Some will die in captivity from the emotional fall-out of this fact.

Researcher Dr. Sally Boysen of Ohio State University has worked with chimpanzees to gain insight into their understanding of numbers and the division of the world around them. In an experiment, Sally challenges a chimpanzee in a game. There are two dishes containing a number of treats. One dish has six raisins, for instance, and the other has four. The rule is that whichever dish the chimpanzee chooses goes to its partner in the experiment. Again and again the chimpanzee picks the larger amount.

Even though the chimpanzees realize that the larger amount goes to their partner, they cannot inhibit themselves against choosing the larger number. This is where the experiment takes an interesting turn. When Sally offers numbers instead of treats in the dishes, the chimpanzees are able to pick the smaller number, which then indicates a smaller number of treats to go to their partner, and they can have the larger amount.

Sally believes that through the use of numbers as abstract symbols for real amounts, early humans were freed from this automatic and obviously greedy reaction to real resources. Numbers became a screen through which instinctual behavior might be modified and more complex social strategies could evolve.[33] The numbered thing would become more remote somehow, approached through elaborate steps and more distant from immediate reactions. I think this is why dead baby gorillas have numbers instead of names. They become emotionally removed commodities rather than real and tangible losses.

As Nonesha slowly went up to her mother and new sister, stopping a respectful three feet away to study the curious little force in their mother's cradling embrace, I was glad that Malaika had lived long enough to be named, to be tangible. *Malaika* means "joy" in Swahili. And that is what she is, in innumerable ways.

July 9   I sat on my bench while the sun streamed into the shelter and created an intense but welcome heat, even though it was still only eleven o'clock in the morning. Newly hatched butterflies fluttered and dipped in shafts of light so intense that dust sparkled in them. The sweet and heavy smell of blooming flowers was a liquid background for the droning hum of bees and flies.

The gorillas were sunning out in the yard. Taufiki and Nonesha lay sprawled on the grass and were immobile except for an occasional twitch of the toes or a swipe at an annoying insect. Adhama, as usual, kept watch over his family, sitting with his back against a tree and a stick across his knees. Binami was lying on her back and dangling Malaika from a calloused leathery foot, poking her gently with one finger to make her laugh and swing.

I was thinking again about Malaika's name and how it suited not only her, but also her situation. I had actually begun my study with a distinct disapproval of zoos of any kind. Though it was true that these gorillas were still captive, I had to concede that they really did enjoy this habitat and the security of their relationship as a family. The people who cared for them here really did try to do the right thing for them.

Before I had become involved with these gorillas it had been hard for me to appreciate how complex the issues are and how people who want desperately to do what is right are eventually forced to choose among several evils. When people step in to correct a bad situation, sometimes it makes a permanent difference for the better, sometimes for the worse. One could cynically argue that more often unforeseen events can make a good situation turn bad. I think that life is usually made up of an inconceivably complex weaving of the two.

Such is the story of one gorilla who, like Malaika, started out with a name that reflected happy times. His name was "Buddy."[34]

During the last part of 1932, a passenger by the name of Gertrude Lintz was on a freighter piloted by a Captain Philips, who, despite having been a kindly man by many accounts, had the dubious habit of shipping baby gorillas to the United States for sale. During the voyage that Mrs. Lintz was on, the sailors had gotten drunk and decided to make sport of one such baby gorilla who had been irritating them somehow. As a climax to the revelries one of the drunken sailors threw nitric acid into the baby's face, permanently disfiguring him with a scar on his upper lip, which gave him a perpetual curling sneer. Captain Philips, to his credit, at once fired the sailor responsible.

Mrs. Lintz, saddened by the baby's condition, told Captain Philips that she had a way with animals and would like to buy the baby in the interest of improving his health and his lot. She was able to purchase him and gave him the happy name of Buddy.

Unfortunately, the little gorilla was destined to have further trouble in his life. First, a disgruntled houseboy working for Mrs.

Lintz fed the little gorilla a bottle of poison, which damaged his intestines. Mrs. Lintz, with a swiftness that would have impressed the good Captain Philips, got rid of the houseboy and began the long, painstaking task of nursing young Buddy back to health.

Buddy did eventually regain a semblance of health and as a result grew bigger and stronger. Ultimately this, too, proved to work against him, for as he grew, Mrs. Lintz felt less and less at ease with him and began to be unable to control his growing assertions of dominance as a maturing male gorilla. In 1937 a grieving Mrs. Lintz reluctantly sold Buddy to the Barnum and Bailey Circus for $10,000. He was six and a half years old. Buddy's friendly and reassuring name was immediately changed to the more awe-inspiring moniker "Gargantua," and Barnum and Bailey spent $50,000 a year to promote the gorilla through advertising and the press as a brute with "murderous intent," capitalizing on his disfigured curling lip.

In 1947 another publicity blitz focused on Gargantua. The circus had acquired a female gorilla named M'toto, and there was to be a wedding. Fortunately, the circus had the sense to introduce the gorillas slowly through side-by-side cages before the "wedding" took place. Gargantua seemed fond of M'toto and tried to gain her favor by tossing his celery to her. She flung it back in his face. His disfigurement surely made it impossible for him to communicate his real intentions, whether to her or others, and so ended his chance at companionship.

On the last day of the circus's 1949 season, Gargantua passed away. The cause of death was double pneumonia compounded by kidney disease, tuberculosis, and cancer of the lip.

The monster so maligned and hawked by media and circus must surely have dangled from his mother's foot as an infant, just as I watched Malaika doing. Maybe, like many mothers who are able to understand the concept of the future, she had pictured him in her mind's eye as a leader of his people. Yet, despite the best of visions, he ended up on a steamer without her for a land she had never imagined. It may have nagged at Mrs. Lintz that buying baby gorillas wasn't right, but what was she to do? Gargantua himself may have had the best intentions during his short life but had to cope with the facts that those intentions would never be known and that the situation could not be helped. The people in this sad story could not change who they were; they lived within a context in which they had only slight agency, but they tried, as each twist and turn of fate appeared, to do what they thought was right at the time.

I watched Malaika laughing. Days ago I was glad she had a name. Now I hope her name continues to sum up her life. Will people look back and feel we did the wrong things for her? I am almost sure of it, but I cannot escape the layers of complexity that surround us both. All we can do is build the future one layer at a time, stacking up the best of intentions.

July 11   "It was stacked up, in layers! One on top of the other, up and up!"

I stared at Karin, unable to take in what she was saying.

"I can't explain it; it's wondrous! You must come see it for yourself!" She panted, then went into a coughing fit. She lit up a cigarette and headed off in the breakneck pace I had come to understand was her primary means of getting from point A to point B. Down the hall and into the office—I let out a polite but hurried belch to the gorillas as we blew past them—down the service hall and finally to the yard. It was always a little surreal to be on the other side of the glass. In the early morning mist I looked over the top of the hill the gorillas crested each day and saw the shelter some thirty feet beyond. My bench stood empty on the other side of the long, curved viewing glass.

"Here!" Karin ejaculated. She grabbed me by the arm and dragged me to a small, hidden place between two bushes against the habitat wall just outside the shelter. She pointed triumphantly.

There, between the bushes, was a stone structure. It was about eight inches high and about a foot in diameter. The gorillas had loosely stacked about fifty rocks varying in size, one on top of another, to form a geometrical cairn. I looked around the habitat and surmised that the rocks had fallen from the top of the habitat wall, where gardening had been done, and must have been subsequently carried to the spot by one or more gorillas. The implications made me reel.

"That's incredible!" I stuttered.

Karin stood staring at it with a look of vindication, her arms folded and the cigarette dangling coolly from her lips. "I told you! You didn't believe me! There it is."

I took the camera from the pocket of my khaki researcher's vest and snapped several pictures from different angles. I was trying to think of ways we could preserve it. Maybe we could glue it together and put a fence around it until it was dry. Or maybe we could number the stones as if this were an archaeological site and take them to the office to reconstruct them exactly, as though to place them in a museum.

"The stones will be gone by the end of the day." Karin said, as if she were reading my thoughts. "I've found these several times through the years. The gorillas build them without us ever seeing them in the act. They leave them overnight; then they are gone by the end of the next day."

She took a long draw on her cigarette while giving me a hard, squinting look down her nose. I sensed that my disturbing the stones in any way was out of the question. I suddenly found that I was a little ashamed about even wanting to take them away and save them. Again Karin seemed to follow my line of thought.

"It really is a miracle," she said. "Your first thought is to save them, to show them to everyone. Papers could be written! Alert the media!" She trailed off chuckling. She looked thoughtfully at the cairn for a minute and then continued. "The stones have to stay, though. They have to be left alone. We take so much from the gorillas. This is theirs."

I thanked her for letting me see them and took one last lingering look at the stones to fix the image in my mind; then I walked back along the circuitous route to my post on the bench, leaving Karin to prepare the habitat for the gorillas' release into the yard.

I stayed almost all day, watching for the gorillas to go back to the concealed spot and do something. I wanted to see them disperse the stones, or stack more of them. . . . Anything. I never did see any one of them go near it. I did, however, watch them build their day nests using the same brand of architectural skill that had been so evident in the stacked stones. Nest construction is an interesting blend of care and style and varies greatly depending on the individual builder. Adults, infants, and juveniles all practice this behavior, and the elaborateness of the endeavors varies as a result of age, sex, and individual preference.

Early in the study Nonesha had been experimenting with practice nests but normally returned to her mother's nest to sleep. Now that Malaika had been born she would have to begin making the real transition to a day nest of her own making, though it would probably still be right against the side of her mother's.

Taufiki built nests of his own occasionally and was even seen carrying grass to different locations around the habitat for the purpose. His nests, like those of his father and indeed most male gorillas, were hardly more than a raked-up pile of readily available material that they sat on top of. For this reason he probably found Binami's nests far more inviting and sometimes shared his mother's after it had been constructed.

Binami made a nest every day, and hers were by far the most elaborate. She would pick her spot, always in the same general part of the shelter, and turn in a slow, clockwise circle while she scooped hay toward herself and simultaneously kicked out the middle with her feet, forming a bowl. She would stop occasionally to retrieve sticks or long, thin branches, which she would then incorporate into the structure as she turned and gathered in more hay. She would often finish by making a rough wooden rim on the top. Karin had

Nonesha imitates her mother in building a nest

told me that the nests were extremely difficult to dismantle each day and that she had once decided to leave the structures intact to see if Binami would like to use them again. Binami, like wild gorillas, showed her disdain for used nests when she shoved the leftover structure aside the next day and built a new one.

Dian Fossey learned that mountain gorillas built nests in the trees as well as on the ground. During the rainy season the sheltered hollows of tree trunks are preferred, and nests in these locations may be made only of loose soil or moss. These types of nests offered the additional advantage of providing a comfortable place to eat, as the gorillas had only to reach from within the nest to obtain snacks of bark and roots.[35]

A study of the archaic origins of shelter building, conducted by Colin P. Groves and J. Sabater Pi, has determined that in constructing nests, lowland gorillas use grasses, trees (including branches and fresh or dried leaves), shrubs, and vines, many species of which grow at the forest edge. The study also concluded that nest/shelter locations also include tree stumps and fallen logs as well as rocks. Further, gorillas occasionally attempted to build nests under the naturally sheltering overhangs of banks or tree-root formations.[36]

The oldest human shelter ever found was in Olduvai Gorge in Tanzania.[37] It was discovered by the Leakey family, famous for their record of uncovering very early human fossils. Like the oldest of the Leakeys' fossils, the shelter is thought to be about 2 million years old. The shelter is a rough circle of stacked basalt stones. The inside of the circle was practically empty of any kind of archaeological remains, but the periphery of the shelter was littered with food debris and numerous tools. Some scientists believe that the inhabitants used the stacked stones as a support for a construction of animal skins and grass or a protective ring of thorny branches. Others believe the stones were stockpiled to use as missiles in case the inhabitants were attacked by dangerous animals.

Who knows? If the gorillas develop and refine their habit of stacking stones in the shelter, they, too, may eventually accomplish such a marvel made of piles of ordinary stones. Maybe that will give us a clue as to which purpose our ancestors used them for. Perhaps we should be prepared to duck. . . .

August 1   The heart of the summer pulsed slowly and strongly around me as I sat on the bench in the heat. I was scratching notes onto the sheet on my clipboard and trying to watch Taufiki at the same time. It had been a lazy morning without much to note: gorillas coming out into the sunny yard and feeding in their usual way, building day nests in the sun, and the rapidly growing Malaika exploring her mother's body as if it were a mountainous playground.

At around 11:00 A.M. Taufiki had started dragging small, short sticks into the shade of the shelter and carefully stacking them up together. This was obviously stacking for some purpose; it was not the seemingly aesthetic stacking of the rocks Karin had found so many times, or the buildup of some kind of arsenal, but something new that I had not previously seen.

Taufiki roamed the perimeter of the habitat intently, looking this way and that, keeping his eyes peeled for the right material to aid him in his mysterious purpose. At one point he disappeared behind the tall trees at the back and emerged some minutes later with his arm loaded with short pieces of wood. Walking upright in his usual manner he began the long trek to the shelter.

Not until this moment did I notice he was limping slightly. Why would he be doing all this work exactly at a time when his foot seemed to be bothering him? All I could do was watch. He made his way to the shelter, awkwardly balancing his load while trying to hold the toe of his left foot off the ground. This was difficult indeed, because apes are at a significant mechanical disadvantage when they walk upright under the best circumstances. Though their precarious gait looks comical and perhaps as if they are performing an intentional parody of humanness, it is really just a practical matter requiring a lot of physical prowess to pull off. Taufiki had this prowess and was one of the most proficient bipedal walkers among the gorillas. Trying to keep his big toe in the air while performing this strenuous exercise was too much even for him, though. He fell twice and had to gather his wood all over again. His lips were pursed with concentration and frustration. He finally made it to the shelter, however, and dumped the load from his arms as he sat down with a heavy thud and remained motionless except for his heaving chest. After he caught his breath he started picking out pieces from the deposited cache of wood and placed each selection carefully on the pile. When the last piece was balanced on the top, he sat back and looked at his crude construction. The finished product was a foot and a half high and about two feet square.

I thought I would at last get to see its intended purpose—but I was soon disappointed. I had not realized how involved I had become in Taufiki's labors. I had been tense and grimacing as he picked his way across the yard, and I had relaxed as he sat down and began stacking wood. I realized I was once more grimacing as he got up and hobbled slowly to a nearby hawthorn. I wanted him to sit down. Surely his toe was painful, and I was worried that he would fall again.

He closely examined the branches. He reached up for one that caught his eye and delicately drew it by its very tip down to eye level to inspect it more thoroughly. With ginger care he reached back along its length and deftly snapped off a two-inch thorn. He let go of the branch, and it whipped up as if insulted by his molestation. He held the thorn close to his eye and turned it around to inspect it before lunging and mincing his way back to the shelter, his big toe still skyward.

He deposited himself in front of the stacked wood structure and with one smooth motion swung his left foot up and balanced it on the top. Being very flexible, he was able to then lean forward far enough to inspect the offended digit. He examined the undersurface of the toe from only three inches away. His searching eyes zeroed in on one tiny spot, and he brought the thorn to his toe.

He began needling into the flesh of his toe with amazing dexterity. The tiny movements could barely be seen. With the sharp end of the thorn he picked at his toe, changed angles, and picked at it again. He repeated this routine several times and then brought his other hand to his foot. Using his massive fingers like a set of the finest surgical tweezers, he pinched at something where he had been working with the thorn. He brought his tightly squeezed thumb and forefinger close to his face, lining it up with the brightest light. There, in his viselike grip, was a splinter.

Even before this amazing instance, the gorillas had engaged in behaviors intimating a rudimentary knowledge of medicinal practices. Binami had almost incessantly groomed a wound on

Nonesha's head, preventing its infection. When Binami had suffered her own illness, Adhama and Taufiki had tried over and over to get her on her feet and herd her around the habitat, displaying at her several times until she got up and moved to the feeding area. During the same illness, Binami was groomed by her offspring, and in general they showed tender concern for her.

Mountain gorillas provide nursing support in the form of close proximity, as demonstrated by instances of group members remaining by a recuperating gorilla's side. An especially poignant case tells of a silverback sleeping near an old and infirm group member during her illness and death.[38] This same silverback had slowed the pace of his group's movements so that she would not fall behind. A five-year-old gorilla in Dian Fossey's study group groomed her mother's bite wounds (the mother was unable to reach them as they were on the back of her neck) for six weeks until they healed. Interestingly, when nonrelated individuals tried to groom the area, the daughter pushed their hands away. Other apes show this propensity for doctoring their family members as well.

Perhaps the most touching of these altruistic endeavors by apes was after Madam Bee, an old chimpanzee in Jane Goodall's study, was attacked by a war party of northern chimpanzees and left for dead. Her daughter, Honey Bee, stayed by the failing body of the mother, grooming her and brushing the flies from her wounds until, after four days, Madam Bee died.[39] In a similar incident, the brother of a chimpanzee stricken by polio stayed near him constantly, leaving only briefly to eat. This same male defended his ailing brother against the attack of a dominant male, which is an extremely rare intervention and usually rife with consequences.[40]

Goodall points out that such concern for ill family members is not uncommon. However, the caretaking of family members can stretch to uncommon lengths. One chimpanzee brother who had become his sibling's parent when their mother died actually wiped the snot off his brother's face with wadded leaves after sneezing bouts during the cold season.[41] Other examples of the compassion of apes extended to the sick and injured include daughters who climbed trees to gather fruit for their sick mother and mothers attending to the wounded or paralyzed limbs of their offspring, carefully and gently arranging the afflicted bodies of the infants. Apes have also demonstrated the applied use of medicinal plants. *Aspilia pluriseta*, for instance, is known to be used medicinally by chimpanzees; it is not eaten as a regular plant food because it is toxic most of the time. The chimpanzees gather the plant immediately after waking at dawn, when the plant is not toxic—before 8:15 A.M. The chimps ingest leaves of only a certain length and crumple the leaves only slightly before swallowing them whole. Researchers have determined that the medicinal properties of the leaf are effective in killing internal fungi, parasitic worms, bacteria, and certain viruses.[42]

One is led to ponder how much memory must play a role in these medical practices. All of these apes must have remembered how to accomplish these feats of ministration. They must have remembered past illnesses of their own—how it felt to be hurt or ill—and the illnesses of loved ones that many times led to their loss. The memories of infirmity and its treatment must be both personal and ancestral to some degree.

Though memories of the first type are something we obviously share with apes, memories of the latter type are not as easy to prove with examples; however, the examples are there. All over Eurasia there are folktales of ancient, wild, hairy people who come in visions and help people understand how to treat the afflicted.[43] In Tibet and China the mythological figure is said to hold the key to treating mental illness. In the Ural Mountains, the local name for this archetypal repository of medical knowledge is known by everyone. Its name translates to mean "Grandmother's Mother."[44]

I have no doubt that Taufiki used clear memories of his own to construct his "operating table" and "surgical instrument." How much his ideas sprang from an ancient internal source—what we call "instinct"—is intriguing but impossible to know.

I think it is safe to assume that whatever grandmother's mother we shared before we were cousins rather than siblings surely sat down on warm days exactly like this day and extracted splinters from her own horny foot with the grandmother's mother of the tree Taufiki used today.

September 13   There is a cool rumor in the air today of the approaching cold season. It is not yet a vicious rumor, just a cool one, the kind shared over tea with the eyes averted. Earlier in the morning I could see my breath. I had optimistically refused to bring a sweater; taking pity on me, Karin had come out to my post on the bench to insist that I put on her old, tattered black coat.

As I looked at my hands—losing their summer tan—jutting stiffly from the moth-eaten cuffs, I thought just how much I considered this coat a gorilla suit. Not the cheap and grotesque kind worn in the deplorable Hollywood movies of the 1950s and 1960s, but a real gorilla suit, a real honorary bit of clothing, like something you wear when you graduate or when you earn a medal.

I brought my arms up and covered my face with the sleeves. The old and familiar smell of gorillas, like lemons and apples left too long in the sun, mixed with the scent of timothy hay and alfalfa and wrapped itself around the smell of the sweet potatoes Karin had baked for the gorillas this morning. I snuggled into the warmth of those smells and the soft blackness of the coat itself. It was a second skin. A living suit. A gorilla suit.

This suit had seen generations of gorillas: held them as babies, cleaned and prepared the places they lived and slept, looked after them in sickness, baked them birthday cakes, carried their food, carried their bodies. Like a soul, it absorbed the changing seasons and became more of itself without changing at all, except for some damage around the edges. Even the damage was something needed for the wholeness.

I looked around myself and absorbed the changing season. The gorillas were spending the days in the shelter now, where it was warmer. The heaters, with their steady red glow, had been turned on beginning with the first day of September. The gorillas sat contentedly on the big, flat-topped boulders rising a foot out of the hay of the shelter floor. The boulders were actually fake. They were, in fact, a decorative cement material called Gunnite, molded into boulder shapes to hide heaters inside. The gorillas sat or leaned on the Gunnite boulders to enjoy the unbeatable feeling of toasty bottoms and feet. They munched on carrots and raisins while they remained in these satisfying positions.

Leaves on the trees in the habitat were turning gold and red and purple. I remembered having read that leaves do not actually turn colors but have the fall colors in them all the time. What happens is that they lose their green chlorophyll in the autumn and reveal the colors that were there all along.

Through their chlorophyll and other chemically active molecules, plants use the energy of sunlight to synthesize sugars and other carbohydrates. Deciduous plants have many light-absorbing pigments—the ones we see at the end of the year—in their leaves and not just the green chlorophyll "factories," so they can absorb a broad spectrum of light to convert to energy.[45]

What is even more surprising than the existence of all those colors in the leaves is the idea that leaves should be black. For plants to make the most efficient use of sunlight, wherein all wavelengths of sunlight would be absorbed, they would need not only the leaf pigments that catch yellow and blue-green light, but additional pigments that would absorb other colors as well. This would make the leaves black. There is no reason that we know of for plants not to be black, and in fact we would expect efficient black plants to have taken over the earth, but they have not.

Regardless of plants' colors, they are in turn eaten by other organisms, such as gorillas and people, who in turn use the energy of the light of the sun to evolve into all kinds of colors. Gorillas have evolved to be as black as the plants they eat might have been but for a twist of evolution, as black and fragrant and living as Karin's gorilla suit.

October 10   Gorillas are black, and like the theoretical black plants of science, they have not taken over the earth either. At first glance there seems to be no correlation between their color and their dismal situation in the human-dominated world. Admittedly, gorillas do not have the capacity for changing their environment the way that humans do, and for better or worse, their choices over millions of years have led to the fork in their road that may take them to a quicker end.

There is no doubt, however, that they have culture. They make tools; they have complex communication; they are capable of abstraction, humor, learning, and creativity.

So why don't we think of them more as people?

The story of Snowflake might give us some clues.[46] Snowflake's origins are obscure. Like other baby gorillas, he was probably around six months to a year old when he was captured in western Africa. He underwent the typical treacherous challenges of losing his mother and family group and being held for some time in a small cage in a village near the place of his capture, deep in the jungles of his homeland. He was purchased by animal traders and eventually took up permanent residence in the Barcelona zoo in Spain, where he still lives as an aging gorilla. He has fathered several healthy offspring and enjoys the company of the females who have been part of his life for many years. It was the Barcelona zoo that gave him his name: Capito de Nieve. Snowflake.

Snowflake, as we have seen, is typical of every other gorilla born in the wild living out their days in the zoo. He is typical in all ways, that is, but one. He is white.

Snowflake is the only known white gorilla, living or dead. He is not an albino; his eyes are a vivid blue. His skin is a pale pink-ish color, and his hair is as white as new snow—hence his name. He has fascinated zoo-goers for decades now, and people who have seen his picture are immediately arrested by his appearance. They stare into his captivating blue eyes set in the pale peachy face under his shock of white hair and invariably say the same thing: "He looks so human!"

Snowflake has the same features as all other gorillas. His nose is wide and flat; he has brow ridges jutting out over his eyes. His hair grows in the same pattern as that of other gorillas, growing from an inch around his face to several inches on his arms. His chest, hands, and feet are bare except for some normal growth on the tops of his hands and feet. The fact that he is not black is the only difference.

Comments that people have made on examining his picture show that they are more likely to think Snowflake is smarter than the average gorilla, friendlier, more trustworthy, a better father, and more likely to possess the whole range of characteristics that are normally the province of human capacity. The doubt of Ota Benga's humanity based on his color during the World's Fair in St. Louis at the turn of the century is something we can easily look back on with outrage. Will it bring angry cries if we contemplate the possi-bility that the gorillas' fate is affected by the same racism? If we took away their color like the green from autumn leaves, would we find them more beautiful? If we took away their dark, handsome faces to glimpse the pale and ghostly shadows of their spirits, would we find them more human? If gorillas could wear pale masks, would we be more likely to see what they really are beneath their dark surface?

October 31   On my way to the gorilla habitat late in the afternoon I was passed by stray bands of school groups. Made up like demons and heroes, they looked like characters shaken whole from the flat pages of a dusty tome of history and fantasy. I was reminded that this is the time of year for paint and masks and doorways to the past that materialize when the living and dying parts of the year are pulled apart like caramel on an apple.

Halloween, as we celebrate it now, is really a thinly disguised Celtic holiday of old times. Originally called Samhain (SOW-in) in the Gaelic language, it marked the time of year when the spirits of all things were drawn back within themselves to wait in patience and meditation for the spring. The trees withdrew their sap and lost their cloaks of leaves. Plants of the harvest dropped their bounty onto the waiting ground, their seeds making the earth pregnant and expectant. Seeds, the tiny hope of all life, were flung across the thin veil of the abyss into the waiting, and perhaps perilous, unknown.

During this fall festival, the tiny hopes of tribal people were flung toward the veil, too. It was believed that the doors between this world and the world of the ancestors beyond opened at this time of year and allowed opportunities to seek their guidance and strength. It was also a dangerous time. Restless souls could walk the earth in terrifying forms and plague the living. Just as the specters of the departed could bestow their gifts of wisdom or take the living unawares, so the spirits themselves had to be appeased or fooled. Who was living and who had passed on? Were the gifts that were given wisdom, or were they folly? Trick or treat?

As I settled in on my little bench I thought about thinning doorways and about barriers not being what they seemed. I remembered being amazed to hear that glass is a liquid. If old glass is measured, it is always thicker at the bottom because it flows downward over the years.

I glanced through the liquid that separated the gorillas from the world and turned often to watch the continuing stream of costumed schoolchildren pour into the observation area, their riotous laughter and the smells of facepaint and sweets preceding them. They were so entertaining that it was some minutes before I turned my attention to the four gorillas in the habitat.

Four gorillas. I counted. Taufiki, Binami and Malaika, and there at the very back, Nonesha. Not all the gorillas were always in sight; in fact, the habitat had been specifically designed to allow the gorillas places to get some peace from prying eyes. Perhaps the loud festivity of the children and the stream of grotesque masks had forced Adhama to take cover until the whole alarming affair had blown over. I was beginning to realize that this line of reasoning was weak, given his spotless record of duty in protecting his family, when Karin appeared around the corner. She was so pale that I laughed out loud, thinking she had gotten into the spirit of the occasion and used pancake makeup on her face. She looked stricken and did not laugh back; I knew something was wrong.

"Someone at the Visitor's Center just called to tell us that a family stopped in to say how much they enjoyed our big man dressed up in a gorilla suit for Halloween," she whispered flatly, fixing me with a stare of growing panic. I blanched to match her as we both turned to look desperately but unsuccessfully into the habitat again. I saw tiny comets swim across my vision, and there was a roar in my ears. My mouth was dry and my lips stuck together as I turned to Karin.

"Oh my god. Adhama's out," I said, though I did not hear my own voice.

Karin turned and ran like a bullet from a gun to the security office as I began calmly telling groups of people that they must proceed immediately to the exit area and await further instructions. There was no need to panic, but please move in a quick and orderly fashion to the nearest gate. That's it, everybody move, nothing to worry about. Disappointed children and irritated parents began the exodus to the gates, slowly at first, and then with decided purpose as beige zoo trucks with sirens came roaring into the gorilla area with zoo personnel spilling out of them, dart guns and firehose attachments rattling and spilling out like a trick-or-treat bag emptied onto the floor.

Karin came running to where I was still standing and came to a halt beside me, gasping for breath. Her face was even whiter and her lips were purple. Her eyes were bulging from her head. I was worried that she would have a heart attack, but as if to keep death away with clouds of magical smoke she lit up a cigarette with a violently trembling hand.

The veterinarian also came over to where we were standing, a medical emergency kit in one hand and a square fishing-tackle box in the other. Without saying anything, she quickly set both boxes down at our feet.

With expert movements developed over years of life-and-death crises, she opened the tackle box, lifted a dart gun from its protective foam bed, and slid in two huge cylinders each having long, hollow needles on one end and silky puffs of yellow and orange on the other. She lifted four additional darts, with their needles still capped, from the box. She put three in her shirt pocket and one in her teeth.

A security guard had begun to organize the search for Adhama. As he gave orders, the khaki-clad SWAT team began to fan out to look for the lone silverback in a practiced pattern with military precision. Karin told me to go back to the gorilla office and wait until it was over. Though I wanted to stay and help in some way, I knew there was nothing I could do. The more people that were there, the more opportunity there was for something to go wrong, and I certainly did not want Adhama to be hurt.

I slowly and warily wound my way through the trails to the gorilla office, and with each step I braced for the sound of the shot that would signal that Adhama had been surrounded and would soon be subdued. I dreaded hearing it. It was always upsetting and painful for an animal to be captured. I tried to console myself with the thoughts that it would be quick and he would not be killed.

There was a rustling to my left in the bamboo that lined the cement path. This was where the path that Karin took to the feeding place on the wall joined the public path to the gorilla office. When another rustle followed the first I thought I had better speak up and avoid getting darted by someone intercepting me from the tall brush. I carefully parted the bamboo that led to the trail.

"Don't shoot!" I said with a nervous smile. I looked into Adhama's face nine feet away.

The realization washed over me that I was in worse danger of peeing my pants than being darted. Adhama had broken off several stalks of bamboo from the lush growth on the trail to the feeding wall and was sitting silently except for his loud munching. He regarded me for a moment over the wad of bamboo leaves sticking out of his mouth, then resumed chewing with gusto.

I slid down carefully to my knees and watched him. With a quick motion he put his hand around one of the bamboo stalks beside him and, holding one end in his teeth, stripped off another huge handful of leaves. He let the naked stick fall into his lap as he crammed the green, leafy mass into his mouth and gave a deep contentment grunt. He flexed and stretched his toes and looked around. He seemed happy. It was clear he had no plans for moving, so I relaxed into the rhythm of his chewing, grunting, and plucking.

I looked away and down to the cement path below my knees, the path that led from here to any point I wanted to go in the world. I thought about how I took for granted my ability to go where I liked whenever I liked. I passed this same stand of bamboo every day. It meant nothing to me. It was a wild paradise for my friend. I watched him in his tiny paradise. Bamboo . . . what more could anyone want? Time ticked by and I knew I had to do something, but the idea of calling out to the people looking for him made me feel like a traitor. Even though he was absorbed in his salad bar at the moment, I knew I was in danger. Though gorillas are normally gentle, they can lash out if they feel cornered. I was as scared as I was content to stay with him as long as I could. After only a few moments my decision was made for me as I heard stealthy footsteps approaching the turn in the path.

"He's here," I said over my shoulder after I had stood up and begun walking away backward. The guard turned the corner, and

I pointed to the opening in the bamboo curtain. He called to the veterinarian. It was a matter of seconds before the whole team had assembled around the jagged opening in the brush. I stood a safer twenty feet away and looked on as the veterinarian knelt where I had been a few minutes before and motioned for someone to part the tall stalks of bamboo. Slowly, the foliage was bent over from either side.

There was a flurry of motion as the veterinarian slowly pushed her red hair behind her ear and took aim. Adhama jumped up and tried to run away down the path. She squeezed the trigger of the dart gun once, then again, and two short pops intruded on the crisp fall air. She then jumped to her feet. Everyone ran to a safe distance. It was only a matter of time now. We waited silently.

The furious rustling in the bamboo became slower and slower until there was one last feeble tremor and all was still. Adhama had lost consciousness.

With halting steps, the team moved closer to part the screen of growth once more. From where I was standing I could see Adhama's inert legs and his enormous belly lifting and falling with his steady breath. I lost sight of him as the team surrounded his still form and began maneuvering him onto a special stretcher. After several minutes, during which the veterinarian checked his blood pressure and listened to his heart, everyone emerged from the brush shaking with the strain of carrying Adhama's dead weight and with the adrenaline of the last half hour.

As they carried Adhama past me into the gorilla office, his hand slipped from his chest and slid off the side of the platform, dangling limply a few inches from the ground. I stepped between a man and a woman near his flopping arm and took his hand in mine. It was four times the size of my own. As I squeezed his thick fingers I remembered the day I had fed him strawberries and our fingers had touched.

We proceeded through the enormous doors of the hall, through the office, and then deposited the snoring dark mound on top of a huge nest of fresh hay that Karin had hurriedly prepared for him.

When we were all out of the cage and clear of the door, Karin pushed the button on the wall that swung the metal gate shut. There was a sterile-sounding buzz and a sour clang as the automatic locking system sealed Adhama inside once and for all.

Indeed, the doorways to our ancestors had been thin this day.

November 3 Surveying the habitat, I could see that one stout arm of the great oak tree by the wall, the means of Adhama's crafty jailbreak, had been sawed off and left on the ground.

Adhama was clearly no worse for his experience of bamboo, dart guns, and winning the "Best Costume Award" in the zoo's Halloween competition. As he came over the hill he saw the real annual prize, one that the gorillas always looked forward to: leftover pumpkins.

It had been a long tradition for the people who worked with the gorillas to bring in their carved pumpkins three days after Halloween and give them to the expectant gorillas, who loved to eat them but rarely had the opportunity because of the brevity of their appearance on the market each year.

When Adhama saw this golden treasure spread among the hay in the shelter, his body tensed with excitement as he let out a food call to the rest of the family. As if there was a subtle variation on the call that screamed "Pumpkins!" Taufiki and Nonesha bolted down the hill in a race for first prize. Binami, with Malaika on her stomach, kept up as best she could. The ensuing free-for-all had all the appearance and frenetic energy of a Bloomingdale's sale.

The gorillas raced around, looking at each pumpkin, judging it for size, ripeness, color, the leftover seed factor, and, perhaps, artistic merit. Each gorilla had a unique pumpkin-collecting style.

Taufiki was a hysterical sight as he raced around the shelter as fast as he could go. He would hurriedly put a pumpkin under each arm, then one in his mouth; as he became progressively slower

Nonesha examines a pumpkin

under the restrictions of his booty his face would show signs of strain as he struggled to stockpile more pumpkins somewhere on his body. Invariably, he would end up carrying three more: one between his knees and another in each hand. Thus laden, he would mince as quickly as possible under the circumstances to the back of the habitat to gorge in peace. This was always a trying feat of coordination, requiring a delicate balance between speed and muscle control. His temptation was always to go too fast and lose a pumpkin, usually the one between his knees. Hilarity followed as he would screw up his face in consternation and then try to get the pumpkin into its original position without sacrificing any of the others to the cruel mistress of gravity.

Nonesha, by contrast, went for size. Her selection process was not tarnished by greed; in her case it was less a strictly culinary judgment and had more to do with her love of wearing pumpkins. She would put them on her head—unfortunately, she never caught on to the possibilities of using the carved parts as eyeholes—and dance around on two legs with her hands stretched out in front of her like some odd, hairy vegetable playing blind man's buff.

Inevitably, she would lose her balance as a result of not being able to see, and the pumpkin's great weight would finish the job as she reeled to and fro trying to save herself before the inevitable crash. She would struggle to a sitting position, lift the pumpkin up with both hands to make sure everyone was watching, and then start the whole process again. Karin reported that in the past it had taken days for the pumpkin seeds and guts matted to Nonesha's head during these capers to dry up and fall off.

Binami's technique was more dignified than that of her offspring's. Like a woman in a supermarket, she looked over the produce with a discerning eye. With Malaika paying rapt attention to her mother as she demonstrated her mysterious and infallible powers of assessment, Binami would linger over her decision, smelling, tapping, and licking. She would squint critically and hold the grotesquely grinning candidates in the air before choosing two or three small finalists. Certainly these were the tastiest of the bunch. I wondered, given their size, if they were pie pumpkins.

Adhama used none of these methods, as quantity never appealed to him and neither did size. Taste, in the sense of gastronomic palatability, seemed to make no difference to him either. He was looking for taste of a different sort. He liked faces. If Binami's style was that of the discerning supermarket shopper, his was that of the educated art connoisseur. Delicately, he made his way through the gallery of pumpkins, pausing here and there to gently roll them around with his great finger and peer into their orange visages.

One of the pumpkins caught his attention. He sat down and picked it up with one hand to look at it more closely, holding it in front of his own face contemplatively. The gallery-goer was transformed into Hamlet as I realized with delight that someone had carved an intricate gorilla head, perfect in every detail. The pumpkin stared thoughtfully back at him. Having made his final selection, Adhama stood up on two legs and, cradling the pumpkin protectively, awkwardly took it back to his corner and set it on his knees. He peered at it appreciatively, running his hand softly over the pumpkin's low forehead.

It is, perhaps, the picture of the artistically sensitive gorilla, able to stand and walk only with difficulty and without grace but having a capacity to admire and place beauty before appetite, that may have led to the greatest hoax ever perpetrated on the scientific community. It was long held by the intelligentsia of the late nineteenth and early twentieth centuries that the primitive ancestral link to apes that Charles Darwin had posited in his theories on evolution must have been large and clumsy, physically slow, with a great and shining brain bristling with nascent human intelligence. This image still thrives in film today; our ancestors, bushy and hulking, ever wily and crafty, challenge and overcome any obstacle by the use of their burgeoning genius while their hunched bodies ride along as so much meat. How they communicated this brilliance to one another through one prized grunt that is repeated ad nauseam is never explained in these movies. However, the point is that such a specimen is exactly what scientists expected to find in the fossil record, and so they were delighted when amateur paleontologist Richard Dawson found evidence for just this sort of creature. Discovered in a gravel pit in Piltdown Common in southern England, the fossil fragments were christened Piltdown Man.[47]

The fossil included a very large, human-looking cranium (brain case) and a distinctly apelike jaw. Authorities of the day concluded that the specimen was datable by the clues provided by other fossil bones found in conjunction with the ape-man. A Pliocene age of many million years was given to Piltdown Man. The scientists saw everything they expected to see. That is most probably why they did not see what they should have seen.

Firstly, the ramus, or top of the lower jawbone, was broken off, making it impossible to determine how the jaw fit with the rest of the head. A dissenter at the time also pointed out that the gravel pit was a jumble of debris and fossils coming from different time periods, and that the jaw and head might not even come from the same period of time, let alone the same specimen. The dissenter's protests were ignored, and his career as a scientist went the way of the dinosaurs. Meanwhile, textbooks were written, other careers soared, and anybody who was anybody swore fealty to what became known as "The First Englishman." Many other fossil finds both soon and for years after were rejected outright because they did not jibe with the incontrovertible evidence supplied by Piltdown Man: early man was a hulking brute with a big brain.

As evidence continued to mount and fossils dated accurately began to tighten unrelenting screws on the Piltdown find, it was becoming harder and harder to write textbooks around him. In the 1950s, J. S. Weiner of Oxford University took on what was becoming the Piltdown Menace. Starting at the beginning, he waded through example after example of inept reporting, substandard examinations, presumptions, and prejudices that had botched a realistic interpretation of the find. It was clear to him that the details did not add up. He took a deep breath and began to whisper to his colleagues that Piltdown Man might be a fake. Other scientists, relieved at this possibility, joined him in collecting the evidence to prove it.

Close examination showed that the teeth had been filed to make them look more human. Under a microscope the telltale scratch marks were revealed. It was found that Dawson, the original discoverer, had dipped the fossils in potassium bichromate to strengthen them. This had stained the fossils a dark color that made them appear much older than they probably were. Finally, tests using the newest dating technology were run on the bones. Their age was 500 years.

Ultimately, the cranium and lower jaw, filed and broken judiciously, were shown to have been deliberately placed in the pit in close proximity as a hoax. Some scientists think that the author of the hoax was none other than Arthur Conan Doyle of Sherlock Holmes fame. Perhaps less glamorous but most likely is that Dawson himself was guilty, his other archaeological achievements having also been exposed as being deceptions. However, none of them rivaled Piltdown Man.

In the habitat, Adhama jealously guarded his pumpkin all day. Like the scientists who shielded Piltdown Man from clear analysis, he was reluctant to scratch its surface or probe its contents. Only when the visitors had left the zoo and Karin called him to come in did he take the first bite of the edge of its rim. Like the scientists of the last century, what he found on the surface matched his expectations. Unlike those scientists, I doubted he would be disappointed by the secrets it held inside.

December 23 The fresh snowfall made the 32 degrees feel even colder, and a wind whipping through the shelter made the weather almost unbearable. It was exactly two days after winter solstice and two days before Christmas. I sat huddled by the heater set thoughtfully next to the bench; Karin had taken it out of the storage closet where it went to rest each spring. It was a happy reunion with an old companion as its orange coils glowed faithfully and its metallic rattle complained of the aches and pains of age as it kept me company in the shelter.

Karin had also insisted that I take a spare snowsuit after I had suffered hypothermia last week. I had stayed out several hours and lost both my equilibrium and the use of my hands. I had had to stay in the gorilla office after my research shift, my feet in a plastic tub of hot water and a bottomless mug of citrus spiced tea in my hands. After I recovered from the disorientation, I was cheered by watching the gorillas in their evening routine. The smell of fresh hay mixed with the smell of burning dust as the office heaters greeted a new winter season for the first time and the tiny particles of the passing summer were now saying goodbye in the form of tiny sparks on the hot steel. Karin pecked out the daily keeper's report on an ancient Smith-Corona, pausing from time to time to adjust the protesting ribbon and refill my steaming mug.

Today, though, besides my trusty Thermos, the heater, and my love for the gorillas, there was no comfort such as that offered by the sights and sounds of the office after-hours. These days in the cold were always unique. I could go all day without seeing a human being. As I focused on the gorillas, time seemed to stand still. It was hypnotic. I would forget where I was. In hindsight I was sure that it was this timelessness that had led me to forget how long I had been in the cold several days ago.

I closed my eyes and thought of sitting in front of a roaring fire in an overstuffed chair and merely reading about gorillas in sensual decadence. I opened my eyes and saw the trees all around, covered with snow. No. This was better.

Karin, seemingly oblivious to the cold, came into the habitat wearing her tattered black coat and her threads of gloves, carrying the gorillas' metal food bucket and a stepladder. She smiled and waved as she came closer, though I could barely see her face through the clouds of steam coming at intervals from her mouth. She walked straight to one of the little evergreen trees outside the shelter and placed the stepladder beside it. Hanging the bucket from the crook of her arm, she climbed to the top of the ladder and brushed the snow off the branches of the little tree.

When she was satisfied with the results of her dusting, she began taking items from the bucket and putting them in the forks of the branches. There were oranges with cloves stuck into their rinds, pinecones filled with peanut butter, bananas studded with carob bits, popcorn balls made with molasses, rings made of raw beets, and baked yams. When these wonderful ornaments had all found places on the tree, she strung long, thin apple peels among the branches and then poked yogurt-covered pretzels onto the ends of each one.

When Karin had finished and stepped back from the tree, I took a long look at it and nodded my appreciation. At first I thought that it was a wonderful twist on the modern Christmas tree, but then I realized that what Karin had created was closer to the original truth and that the modern Christmas tree itself was the twist.

Since our beginnings as an arboreal species swinging through the forests, trees have been the focal point of our existence. The living tree, overflowing with bounty like the food treats Karin had placed upon the evergreen, is an archetype of abundance. In one story from Norse mythology that recalls the giving spirit of the tree, Odin hung on the World Tree or Tree of Knowledge for days without food or water as ravens ate his flesh. He was wounded in the side with a javelin. When he could endure no more, the alphabet of runes and their meanings were revealed to him. Even the fruit of knowledge was fruit from a living tree. This archetype turns up again during the revelation story of Genesis, although here the church's early ambivalence toward nature is revealed in the flavor of the story.

This ambivalence was spread far and wide. The Celtic and Gaelic people of Britain and Ireland acknowledged their deep spiritual attachment to trees in rituals that were already so old during early Christian missions into their lands that their origins were a mystery.[48] Trees were associated with the bounty of nature and the magic of rebirth. Yule logs were cut in a ceremonial way and were often carved into the shape of a bearded old man, then festooned with evergreen branches. The log was burned carefully and never completely, so that it could be used to light the Yule log the next year, while the clan feasted on nuts, wild meats, and other such wonderful harvests. The Irish said that the mugna, or yew tree, bore a trinity of fruit: apples, acorns, and nuts. The Welsh Gaels said that the mythical hero Lleu took refuge in an oak after dying, Christlike, from a spear in his side. There he transformed into a marvelous eagle perched in the tops of its branches.

It was relatively easy—given the similar symbolism of trinity, sacrifice, and resurrection—for these old pagan beliefs to continue under a veneer of Christian piety and eventually become absorbed. The church, eager to keep the members it had won, allowed the jumble of traditions to become incorporated into Christian practices. Even today people rarely ask obvious questions about what pine trees, reindeer, a bearded old man, and roasted chestnuts have to do with a wonderful man from the desert. Somehow it just seems right to them, and in a way it is. It is all about rebirth . . . and hope.

The church's ambivalence and later antagonism toward nature made the winning of members far more difficult in other parts of the world, where people continued to have a direct and holy connection with nature for survival. American Indians, for instance, could not understand the evolving Christian battle with the earth, though they understood the idea of sacrifice and goodness. Even the image of Christ on the cross was not foreign to some tribes. The people of the Plains had long held sun-dance ceremonies where the men were painfully affixed to a common central pole of wood and danced around it for days without water or food. Sometimes visions would come to them as they sacrificed themselves for their people's well-being. Though Christianity eventually became an undeniable force in the lives of American Indians, the sun dance thankfully still goes on, a very personal and immediate example of sacrifice and rebirth.[49]

In Africa, European Christian missionaries were heartened when they first heard that gorilla-hunting people of the rainforest thought of the forest as their father. It would be, they thought, a simple matter to convince them that their father was in heaven: same father, different location. What the missionaries did not understand was that a punishing father was nothing like a benevolent and providing one, no matter where he was located.[50] Anyone who knew gorillas knew that. And I had to agree.

So I sat, on Indian land, looking at a Celtic Yule log full of bounty that had existed for millions of years, during a season of celebration dedicated to a man who lived in the Middle East two thousand years ago, waiting for a special kind of African people who had made their way here through the help of an ethnic composite with arms akimbo. I felt like giving everyone the benefit of the doubt. I felt a peculiar peace of mind that comes when you realize there is nothing to fight about.

The gorillas, who seem to have this kind of peace much of the time, came over the hill and spotted the wonderful tree, singular and beckoning in the unbroken blanket of snow. I don't believe they gave much thought to where such a tree came from, the wars fought and the cultures lost, the contradictions of history that brought it here. For them it was here all the time.

January 1 Another bitter day awaited me as I walked down the meandering snow-covered paths to the gorillas' home. My heavy snowmobiling boots dragged with each step, and their rubbery stubbing sound punctuated the embarrassing *shoop-shoop* of the too-big nylon snowsuit that I had now adopted. I peered out from behind the long wool scarf wrapped around my head; everything in my limited view was framed by wool pills and fuzz.

The paths were deserted except for the rare bundled visitor leaning in utter quietude on the rails, watching animals that they clearly had some special attachment to. I thought again about the study that had concluded that the average length of stay at the gorilla exhibit by any visitor was five seconds. I admired these people who came to visit in the winter. Sometimes I would pass them at their posts on the way in and again on the way out.

This day of all days seemed to demonstrate their commitment. Not only was it awfully cold in the literal sense, but it was the first day of the year and early enough in the morning that most people who went to sleep last year to float on champagne dreams had not even opened their eyes on this year's first headache. Cold and fatigue notwithstanding, I would not miss this day at the zoo for the world. It was Adhama's and Binami's birthday.

Not really, of course. It was standard zoo practice for wild-captured infants, their real birthdates unknown, to have a year automatically added to their ages on January 1st. This year was no exception, and I knew that a cheerful scene at the gorilla office this morning would be all I would have to keep the gloom away during the rest of the cold and cutting-steel day.

I went through the massive double doors as they protested even louder than usual against the chill, and then down the hall to the office. As I opened the doors to the office I saw Karin holding aloft a huge triple-layer carrot cake with raisins. The office smelled like cream cheese frosting and the heady winter spices of cinnamon, nutmeg, and ginger. Karin turned off the light, and a flickering glow from a single enormous pillar candle, protruding from the center of the cake like the leaning tower of Pisa, illuminated the happy faces of the gorillas and danced from their dark eyes.

As if I had stepped through the door on cue, Karin and I both began an off-key but heartfelt "Happy Birthday." When the song was finished I clapped for the gorillas' good health as Karin blew out the candle. I flipped on the lights and offered to get a knife from the kitchen drawer.

"Oh, no," she said, "they like to gouge out great heaps with their hands, you know!"

And so she went around to each gorilla and let them all scoop a messy handful, though many times she had to move the cake away from them while laughingly scolding them for trying to be stingy as they tried to put another hand in the cake. I got out a stack of birthday cards I had made them. Karin looked at the cards and made a sad face at me.

"They will eat those. . . ."

"I'm hoping so," I said, trying to reassure her.

I stepped up to Adhama's window grate. His handsome face was smeared with frosting, and he was absorbed in trying to make his tongue reach dollops of the creamy treat by his nose. I held out the card. He looked at it a minute before taking it with his frosting-covered fingers. It did not look appetizing to him next to the cake, I'm sure, but he politely indulged me. When he opened it he found a picture of himself done in a mosaic of raisins, walnuts, and Minute Rice. Karin clapped and whooped, throwing her head back.

"That is just smashing!" She was clearly delighted, and so was Adhama.

"I used school paste so it wouldn't make them sick," I said, suddenly feeling proud and bashful. I passed around the rest of the cards and did not feel one bit bad that three hours of work was obliterated in a mere few seconds.

Reluctantly, as the last of the treats were passed out and a round of cherry Gatorade had washed them all down, I zipped up my suit and forlornly made my way out to the lonely bench, taking a leftover bit of carrot cake with me. The gorillas were as reluctant as I to come out into the unrelenting will of winter. They made their way to the shelter, stopping often to look back, hoping maybe that Karin would call them for more cake. Though I shared their hopes, I knew we were stuck out here together for the rest of the day.

To pass the time, I started another chorus of "Happy Birthday." Adhama came over and sat close to me. I leaned forward to finish the last lines, and he grunted along good-naturedly. My breath made a gray cloud on the glass, and he leaned around it to look at me. I was about to wipe it clean when I decided to make a picture in it. I took my glove off and traced a gorilla face in it, blowing a couple more times to make a place to draw ears. Adhama looked at it. I had the sudden hope that he would copy my actions. But he did not. In fact, he seemed bored.

I put my hand back in my gloves, and that is where they stayed. As the day wore on, it seemed to grow colder rather than warm up.

The gorillas did not move unless they had to; Binami went to the edge of the ceiling heaters' range to squat and pee once before racing back to the nest she had built right against the boulder heater. That was it. I looked at my watch. I had been out here only an hour. Time flies when you're having a good time. How true.

Perhaps that is why we do not know what time is—because we are always having some *kind* of time. Of course, if you ask others, they think they know what time is, until you ask for a definition. They are likely to tell you they know what time is, they just can't explain it. This is an honest assessment of the facts.

Isaac Newton pictured time as being absolute, true, and mathematical. He declared that time by nature flows evenly without relation to anything external.[51] It has become clear, with the help of Albert Einstein and others, that time is actually a subjective and relative thing. For instance, years, as we tend to think of them, are virtually meaningless. The longest year on record was 46 B.C., and the shortest A.D. 1582. These discrepancies reflect human influence. Julius Caesar added two months to the year and twenty-three days to February on the calendar he adopted in 46 B.C. to "correct" the accumulated slippage of the Egyptian calendar. In A.D. 1582, Pope Gregory announced that the 5th of October would in fact be the 15th (why he felt this was an important decree is murky).[52]

Beyond the human perception of the year, the rotation of the earth does not offer a precise standard of measurement, as gravitational pulls and surface action affect the speed of rotation. In the interest of accuracy, "leap seconds" are inserted to a chosen day at midnight when clocks in national laboratories all over the world agree that it is warranted. Of course, how do they know that the second inserted is really a second?

Since 1967, the International Commission on Weights and Measurements has defined the second based on the time it takes an electron to spin on its own axis inside an atom of cesium. This is the so-called atomic clock, and it can measure a second to an accuracy of thirteen decimal places. The shortest measure of time attempted at present (whatever that means!) is the decay of elementary particles, which is an unimaginably brief portion of a second.[53] This is against the backdrop of the estimated age of the universe—16 billion years. It is enough to make your brain as molten as time itself.

Perhaps, as evidenced in the melting clocks of Salvador Dali's surrealist paintings, the poetry of imagery is the only way we can contemplate such conundrums as time comfortably.

I looked at my watch. Another hour had inched by. I wondered if it was cold enough on my bench to alter the subatomic properties of cesium. Certainly if there were any in my butt they had slowed down. I noticed that Adhama, though still near me, had shifted his position so that his wide back was turned to me. He was moving, but I could not see what he was doing. I peeked around his hairy shoulder to where his hands were on the ground. To pass the time, he had swept away the hay in front of him so that the ground was bare; he had a tiny stick in his hand and was focused completely on the space between his legs. He had scratched out a design. It looked like a star.

January 31   It was my turn to hear "Happy Birthday." I stood in the warm gorilla office and looked around at my friends. Karin was the only one singing the words, but I was moved to tears when the gorillas all grunted at least once during the song. Though I was not the recipient of a triple-decker carrot cake, I got something better. Held ceremoniously in Karin's hand was the auspicious birthday treat that had been sung and grunted over so heartily. It was a baked apple.

"It is so weird, you know. Adhama left this. Wouldn't eat it. It's his favorite. . . ." Karin looked at the apple thoughtfully.

"I'd like to think he knew it was my birthday," I said sheepishly. "I guess that's a crazy thought."

"Oh, I wouldn't say that. I never say never when it comes to gorillas. He could have meant it for you." She said this with a frank gentleness that was uncustomary.

"I will enjoy it with that very assumption." I looked at Adhama. He looked at the apple and grunted loudly, insistently.

"Do you want it back?" I asked. I walked over to his window

grate. "It's the custom in some places to *give* presents on your birthday. Here, have it back. It's your favorite." I smiled at him.

He took a blade of hay from the floor and shoved it through the grate and fixed me with a stare I had not seen before. I took the blade of grass. I thanked him and rubbed his finger tenderly with my own.

"Come here!" Karin said loudly. I thought she would give me another lecture on touching gorillas and the resulting bloody stumps. I had sat through more than one of these diatribes; usually she forgot them as soon as she ran out of air, which, given the presumably thick tar coating of her shriveled lungs, was amazingly slow in coming. I was relieved that instead of being angry she produced a box, about a foot square, with a flourish.

"Open it now!" she barked, then smiled.

I slowly untied the sparkling white ribbon and slowly lifted the lid. It was Adhama's drawing.

"My God! I can't believe it! How did you get it out of the ground?" I was breathless with excitement.

"It was a pain in the ass, I can tell you. I went through three cigarettes getting that frigging thing out in one piece. Well, I knew after the stacked rocks, you'd really like it."

"This belongs to the gorillas, too, though. I really don't want to take anything from them. . . ." I trailed off into silence.

"No. It's not like that. You belong to them. You are one of them. Happy birthday," she said.

I did not know what to say. I gave her a warm and lingering hug. My friend. My friends.

It was the best birthday of my life.

February 26 As I set my bag and clipboard on the cold bench in preparation for another day of taking notes, I noticed that the heater was gone from its usual place and then noticed that a folded slip of paper had been tacked to the seat of the bench. I took off my gloves and unpinned it carefully. A sense of dread crept over me. I unfolded the paper as I felt time stand still in the way we all do when we know somehow that our lives are about to change in ways we would rather they did not.

I read the lines. In a hurried scrawl, Karin had written that Adhama was ill and there were to be no observations today. I was to go directly to the office.

I walked numbly to the office with a heavy iron taste in my mouth. As I walked down the hall and reached for the door, I felt the sensation that my body was not mine and it was some stranger's hand that turned the doorknob in slow motion and pushed the door open to the waiting scene.

The veterinarian and Karin knelt on the floor in front of the spare room that was used for weighing the gorillas and for examinations. They looked up at me as I came into the room. Everything was silent, the thick kind of silence that seeps into your soul and makes you feel like you are drowning.

Adhama was on his knees, leaning forward with his shaggy head cradled in his hands. He was staring at the floor, breathing slowly. He did not move.

"What's wrong? What's wrong with him?" I asked softly. I sounded like a child to myself.

"We can't tell. We've given him pain medication, but he won't move." Karin turned her face away. "It doesn't look good, I'm afraid." She said this in a way that was not brave, but accented the fact that she was really afraid.

"What can I do?" I could hear the note of panic creeping into my own voice.

"Go home."

"I won't."

Karin did not have the energy or desire to ask me again. I put down my bag and clipboard and sat down with my back against the cool cement wall, facing my inert friend with my shoulder against Karin's. The veterinarian finished writing some instructions on a vial of pain pills, set them quietly on the desk, and left without ever having said a word. Karin and I silently agreed there was nothing to say. After thirty minutes, Karin got up to grab her cigarettes off the desk and went into the hall, closing the door behind her.

I was alone with Adhama.

Neither of us moved. I felt like a fossil, or one of those people you hear about who is buried alive. Finally, I spoke to him.

"I don't know the songs of the gorilla nation," I said quietly. "I don't know your songs for healing, or setting free, or comfort. . . ." My eyes blurred and came into focus again as I looked down at my feet and heavy tears fell in slow motion, making three drops on my boot.

"I would do anything for you. You have loved me so well," I choked out, "and so differently." He remained still.

I tried to think of some way I could make him understand. I reached forward and took a piece of straw from where I was sitting and stretched my arm into the cage. I set the blade of straw near his elbow where he could see it and held on to it a moment before letting go. Then I sang him "You Are My Sunshine"—the only song I could think of. I ended with

You'll never know dear
How much I love you
Please don't take
Adhama
Away. . . .

Though it was not a song of the gorilla nation, like most songs, it says the same things that gorilla songs say. You are special; you are needed. You are a part of this world. Don't leave.

"Please don't die," I whispered. He made a tiny motion, turning his head the slightest bit so he could just see me with one eye. We looked at each other for a long time. He grunted once, so softly I could barely hear it, and then resumed his focus on the floor inches from his face.

I stayed until late afternoon, just to be with him. When I left, I hugged Karin quickly without looking at her so we would not cry. I said goodbye to her, and then to Taufiki, Nonesha, Binami, and Malaika, who were all holding silent vigil in their own night rooms. None of them looked at me.

I walked out into the winter's early night, closing the huge double doors behind me. I stood hanging on to the worn rings that served as their handles and looked up at the stars—the stars that Adhama and Karin had watched together from the hill in the yard. The stars that looked down on my birth and the births of a million ancestors, the stars and ancestors we shared, each one like the star Adhama drew. Their ceaseless light was a comforting rhyme about eternity on a night that was dark both in the sky and in my spirit.

February 27 The sunlight that reaches the earth each day started out in the sun's core 30,000 years ago, about the time of the last Neandertal people, slowly inching its way through crowds of atoms to finally be shot out to make its eight-minute trip to the earth. The light that started out millennia before Adhama's birth and raced to meet him this morning would find that he is gone.

I knew this when Karin was waiting for me outside the shelter that had become my second home. She looked at me with a sadness and compassion that pierced me, tore at me, crushed my chest.

"Dawn," she said, "I'm so glad to see you . . ." and like the first time she had said it to me, on the occasion of our first meeting, I knew that she really saw me. I buried my face in the timeless smell of her tattered black jacket and fought tears until my throat ached. Please make it not true, I thought. Please make it not true. Not him. He had suffered too long and he was still so good. I could not take it. It would mean there was no justice.

Karin and I walked to the office and went into the spare room, now empty except for a scattered pile of straw and a few wrappers for sterile gauze and needles. There was a full I.V. bag against the wall where I had been sitting yesterday. We sat next to the scooped-out nest where Adhama had died an hour ago.

"He was doing a little better, I thought. He was looking around a little and had some Gatorade. Then he got a sort of frightened stare and his elbows gave way. He was in seizure and I called the vet right away. She was here with a team in a minute and a half. They tried to start an I.V. and began artificial respiration. They put electrodes on his chest." Karin's voice was quivering as she finished, "I stared at that goddamned green line on the monitor. His heart just wouldn't start beating again." She put her hand over her face and stood frozen before saying, "No. No. No" over and over again. She was sobbing.

It was devastating. Like the gorillas left behind, we felt we had lost the kindest of leaders; we were now lost.

I bent over the desk. Adhama's chart was open and staring in dumb silence as if it, too, were dead, and someone had forgotten to close its eyes. #33797—"*Adhama*"—Male gorilla—wild caught infant—Cameroon—deceased February 27, 10:00 A.M.—heart failure. It occurred to me that this was the first time I had really thought about where he had come from and how far he had been taken.

"I want to take him home," I said.

Karin sniffed deeply and said, "You can't," flatly from behind her hand.

"Why not? He's dead, for god's sake. What more can anyone squeeze out of him?" I was angry now. It felt better than devastation.

"Because the Harrington Zoo still owns him." Her soft answer waved away my wrath like thin smoke, and I felt only devastation again.

Karin took my hand and pulled it to her. She reached into the pocket of her black coat with her wet hand and brought it out in a tight ball. She placed it in my hand and opened it. A warm, damp ball of black and silver hair was left behind as she put her hand back to her face.

"They shaved this off of his arm while they were trying to get the I.V. started." She began crying again. "Please take him home where he belongs."

March 4   They say that to a first approximation all of the species that have ever lived on earth are now extinct. What they mean when they say this is that the 10 million to 15 million species that are living on the earth today make up only 0.1 percent of all the life that has ever inhabited the planet. This is because it is the natural way of things for species to thrive for a time and then become extinct. It will happen to the gorillas, and it will happen to us. We cannot stop the eventual extinction of any one species any more than you can keep one of its members from dying. Adhama died, I will die; all of us will.

Gorillas find this as hard to accept as human people do. Adhama's death had a profound effect on the group. Before he had been taken away, the gorillas were allowed to see him. They went to his body and poked at him, and Binami would not leave his side until she was forced to. The following day, the gorillas were crying and wanted to be near the people they knew well. Taufiki cried softly by himself after the others had gone out to the habitat, and before he finally joined them, he stopped and smelled the place Adhama had died. For several days he continued to stop and smell the place every time he passed it. A week after the death, Nonesha stopped at the place where Adhama had died. She picked up some dried vegetation that was lying where Adhama's head had fallen and sniffed it, then put it down in the same place. Displays did little to alleviate their grief.

It is no different in the wild. Dian Fossey described in detail one group's reaction to the death of one of its females. A male led the group in wild displays of screaming, charging, and tearing at vegetation, occasionally hitting the body itself. The male dragged the body several meters. The gorillas had nested near the body that night.

Adhama

Mountain gorillas, perhaps trying to spare their families this agony, take pains to conceal themselves in the hollow boles of trees when they are dying, and one gorilla of Fossey's study was found in a tree bole nearly covered with vines. Indigenous Africans say that gorillas cover their comrades' corpses with heaps of leaves and loose stones that they collect for the purpose. Thus cared for, their comrades' bodies once again return to the soul of the forest that will see the rise of future generations.

As I stood deep in the rainforest of Cameroon, near the place where Adhama had been captured many years ago, I thought about this fact. The heat was sweltering and moistly clinging like a great soft womb. The birds were singing, and the light of the earth's green emeralds sprouted and grew in a joyful riot, reaching for my pain and lifting it toward the sun.

I kneeled on the ground and cleared away a small patch, scooping out a shallow hole. I reached into my pocket and withdrew the only part of Adhama that would ever see his home again. I lifted it to my face and breathed in deeply the last of his scent. I held it in my lungs and then blew it up to the sky. I placed his hair in the hole. From its special box, I carefully lifted the hardened dirt bearing Adhama's drawing of a star and used it to cover what was left of Adhama, now resting in the earth.

After patting the soil down I took out a sprig of bamboo from the tree he had eaten during his last escape. I suddenly realized that I had been alone with him the last time he was free and alive and now again as he was free and gone away. He would never be behind bars again.

I was humming a tune, the melody of the last song I had sung to Adhama. I had been right that our songs are the same. They are not only about loss, they are about living, and as we sing the songs we are free. . . . We will not live a death, we will live a life.

I stood and walked a few feet, then turned for one last look at the small place on the ground that would soon disappear under new growth. Somewhere far in the distance I heard a reassuring gorilla grunt and the unmistakable sound of an infant. It was learning the songs of the gorilla nation.

More about Gorillas An Ethological and
Evolutionary Primer on the Anthropoid Apes

In the connective narrative that science has provided us, the Ceno-
zoic era (from 65 million years ago, or M.Y.A., to the present) is com-
monly accepted as the beginning part of the ape/human story. The
evolutionary path of the primates that eventually became humans
and their close relatives occurred over its span.

At the end of the Mesozoic, after the mass extinction of rep-
tiles terminating that era, mammals began the long process of diver-
sification to fill the vacant ecological niches then made available.
During the Paleocene epoch (ca. 65–52 M.Y.A.), the earliest ancestral primates begin to appear. Designated as Plesiadapiformes, these
archaic primate forms lack many features used as criteria for categorization in the primate order. The Plesiadapiformes inhabited the
holarctic range (throughout the northern hemisphere) on the un-
separated continents of North America and Eurasia.[1]

During the Eocene epoch (ca. 53–36 M.Y.A.), an extensive ar-
ray of fossil primates possessing definitive primate features began
to be seen. The dispersal of these primates covered North America,
Europe, and Asia. Some of these forms were probably related to
modern lemurs and lorises, and others to the tarsier.[2]

Specimens ancestral to the anthropoids may appear during
this epoch, but it is in the Oligocene that more solid evidence takes
shape. The Oligocene (ca. 37–22.5 M.Y.A.) was a crucial epoch for
the developing primate order. Primates including New and Old
World monkeys, apes, and humans can trace their emerging ancestry to this epoch, though not with optimal precision. The debate continues (and will probably continue forever) as to the correct
lines of descent—and the circumstances brought to bear on these
lines—surmised from the fossil record. It is known that the Platyrrhini (New World monkeys) separated from the rest of the primate
order at this time. Many scientists believe that this was due strictly

to the disconnection of what is now South America from the original supercontinent to which it was attached, while others believe that other evolutionary pathways were taken even before this continental detachment.[3]

The next epoch, the Miocene, is of particular importance, spanning ca. 22.5–5 M.Y.A. The Platyrrhini were separated from the anthropoids that continued on to form the lines of apes and humans. The complicated history of our origins and those of the anthropoid apes takes shape at this time. A radiation of many anthropoid forms occurred across Asia, Africa, and Europe during the Miocene.

The Miocene ape fossils obtained to date are classified as hominoid, which means "humanlike" and is used to designate all apes and humans, extinct and contemporary, from the time of the divergence between New World monkeys and all subsequent Old World forms. The similar term *hominid* refers to the successful form of Hominidae, which includes all bipedal hominoids. Traditionally, this definition includes the caveat "back to the time of divergence from African great apes."[4] This caveat illustrates the assumption that walking on two legs is a recent adaptive strategy and part of what makes a human being distinct. Anthropoid apes, however, also engage in bipedality in differing degrees of habituality (from only occasional bipedality to total bipedality), and their bipedalism, among other surprising behaviors, brings to the fore the question of biological/behavioral continuums and the arbitrary nature of nomenclature and categorization in general.

According to genetic evidence, the African anthropoid apes split completely and decisively from the ultimately human line around 5 M.Y.A.[5] This is an extremely recent date when viewed from a broad evolutionary standpoint and one that intimates a persisting close connection. Yet in that time, the very success of the emergent *Homo* line and perhaps the early evolutionary choices of the African anthropoid apes have driven modern ape populations down to paltry numbers and very limited ranges.

The three subspecies of gorilla recognized currently were at some point probably part of an original, contiguous group that ranged across the north of the Congo-Ubangi-Uele river system from the Atlantic coast to the Rift Valley of west-central Africa. During interpluvial periods of the Ice Age, their habitats shrank, forming islands of original habitat and effectively isolating populations that eventually gave rise to the extant three species. *Gorilla gorilla gorilla*, the Western Lowland gorilla, inhabits west-central Africa, including Cameroon, Rio Muni, Congo, the Central African Republic, and Gabon. The Eastern Lowland gorilla, *G. gorilla graueri*, survives in isolated pockets in eastern Zaire and nearby highland areas near the border of Zaire and Rwanda. The mountain gorilla, *G. gorilla beringei*, populates the Virunga volcanoes of Zaire, Rwanda, and Uganda. The differences in subspecies are primarily those of size;[6] hair color, density, and length;[7] and varying facial widths.[8]

Western and Eastern Lowland gorillas (*G. g. gorilla* and *G. g. graueri*, respectively) inhabit the warm and humid equatorial rain forest, gravitating to the rich understory where sunlight actually penetrates the thick canopy.[9] Such understory comprises woody herbaceous plants. Where logging and cultivation have taken place, the secondary forest is extensively utilized by the gorillas and in fact may benefit them and lead to increasing numbers.[10] Where completely open fields occur, they surround the lowland gorillas' trails; lowland gorillas routinely travel through these fields but prefer the cover of the rainforest and dwell primarily in the zone at the edge of the forest.[11]

The mountain gorilla (*G. g. beringei*) inhabits the montane forests and Afro-alpine meadows occurring on the extinct volcanoes of the Parc de Volcans.[12] Their ranges throughout the montane forests include relatively flat "saddle zones" between volcanoes, interspersed with hills and ridges, and *Veronia* zones on lower slopes in which *Veronia* trees predominate. The nettle zone occurs both in the saddle and on the lower volcanic slopes in a dense belt of foliage, and bamboo grows in limited areas of saddle regions. There are also brush zones in which shorter, denser fruit bushes predominate, and giant lobelia zones. The Afro-alpine areas encompass the highest portions of the summits and comprise open grasslands and lichen-covered meadows.

In terms of their physical characteristics, gorillas in general have a robust and muscular build, broad hands and feet with thick digits, muscular lips and large teeth, and darker to lighter skin ranging from blue-black to brownish gray.[13]

The hair of the saddle area of the shoulders, back, and hips is white, gray, or silver in mature males of all three varieties.[14] Gorillas are sexually dimorphic, which means that, among other sex-linked differences, males are larger and stronger than females in all three subspecies.

Not only do gorillas and humans share many physical characteristics—differing significantly only in size and adjustments for walking primarily on all fours—but as research continues, it is clear that gorillas engage in many cultural behaviors that were once thought to be the exclusive domain of the human animal. As we have seen throughout the book, gorillas engage in a whole list of cultural behaviors such as

> gathering and hunting
> using tools
> transporting objects
> having primary relationships of teaching and support
> > between mother and young
> imitating, learning, and creating
> building nests and shelters
> having intermediary roles (males act as go-betweens on the
> > group's behalf)
> caring for the sick or injured
> playing
> participating in group rituals
> having complex vocal and gestural communication
> recognizing and mourning death[15]

In gorilla culture, as in any other, the members enact these behaviors, passed down to them through combinations of genetics and instruction, to live successful lives full of meaning in context.

As we know now from unfortunate experience, the preservation or demise of a culture impacts all who must go on impoverished as a result of its knowledge being lost. Ethnologists know this as they scramble to learn from the surviving members of cultures that are soon to die. I felt the desperation of those ethnographers as I wrote this tiny piece of story about a small band of what was once a great nation.

Current estimates place living, wild populations of gorillas at anywhere from 30,000 to 50,000. The most severely endangered are the mountain gorillas: only 620 left alive in the 285-square-mile area that is their dwindling range. Because of their extremely low numbers no mountain gorillas live in captivity; every one is needed in the wild.[16]

Species become extinct for varying reasons. The dinosaurs, for instance, probably met extinction through a complex combination of emerging factors such as climate, disease, decreasing food supply, and perhaps even the unfortunate discovery by nasty little mammals that their eggs tasted good: the same pesky, tree-shrew-type mammals that would eventually give rise to us and the apes. The extinction of the dinosaurs, however, was not the worst or the last mass extinction the earth has seen and will see again.

Regardless of any animal's brilliant success, it will finally leave this place and make room for others. I have heard this inescapable reality cited as an argument against the rights of animals. If, the argument goes, all animals are doomed on the cruel stage of nature, why worry about sustaining them in comfort? We are prolonging the inevitable; this is the law of survival of the fittest.

I obviously disagree. Yet I do not believe that animals have rights—not, at least, in the sense that living things have a right to live without pain. Pain is everywhere, and pain can be a gift. What I do believe is that as humans we have not even begun to reach our capacity for compassion. Our compassion makes us fit—fit for the cruelty and beauty of nature.

Compassion is not contingent on rights any more than the will to breathe is contingent on air. We have a will toward it that ceases only with death. The psychic organ of compassion is as vital to life as any other organ of the body. It would seem ridiculous to the average human or scientist (being one and the same) to study animals with only one eye because the other gave a different point of view that did not match precisely. But that is what we do when we offer science as a view from one eye and emotional narrative as a view from the other.

In reality, keeping both eyes open gives a clearer picture with depth and context and requires the interaction of motion, readjustment, and revision. Amazing things have been recorded in scientific papers, only to be forever buried from the general public, who are seen as uneducated and uninterested. The general public in turn has no idea that such interesting and pertinent things are going on and thus continues to be uneducated if not uninterested.

We are creatures of context. One of the first things I was taught as an anthropology student was the importance of cultural relativism. I took this to heart and decided early on that any work I did would be constructed for sharing with context in mind. I wanted my work to weave science, and stories, and poetry; that is what life does. I knew it would perhaps cost me a traditional educational path and the eventual prizes of professorship, tenure, and the resulting glory. But, as we know, tenure will not prevent death, and as my life goes on and I look back on what I have done, I know I will have lived a good animal life.

I believe there is a shift occurring in the telling of stories in science. I hope there is a shift toward the celebration of the animal within and without—nature, riotous and full of possibility—where it is better to try new things than to be right. Our stories need to try harder to inspire and less to claim authority, because they simply cannot. Only time and nature can know how things really are.

Notes

Prologue

1. Henry A. Ford as quoted by Bourne in *The Gentle Giants*, pp. 133–134.
2. Ibid., p. 131.
3. From an issue of *Popular Mechanics* dating from the 1930s.

Book of Days

1. Schaller, "Behavior of the Mountain Gorilla."
2. Fossey, *Gorillas in the Mist*, p. 67.
3. Ibid., p. 79.
4. Schaller, *The Year of the Gorilla*, p. 131.
5. Ibid., p. 190.
6. Bradford and Blume, *Ota Benga*, p. 97.
7. Gribbin and Cherfas, "Descent of Man—or Ascent of Ape?"
8. See, for example, Tutin and Fernandez, "Insect Eating by Sympatric Lowland Gorillas (*Gorilla g. gorilla*) and Chimpanzees (*Pan t. troglodytes*) in the Lope Reserve, Gabon" and "Gorillas Feeding on Termites in Gabon, West Africa"; Rodgers et al., "Gorilla Diet in the Lope Reserve, Gabon."
9. The formula that shows that the requirement in pounds of food in anthropoid apes is equal to five percent of their body weight in pounds is expressed this way: $^{0.73}$x0.184. A female gorilla can weigh between 225 and 275 pounds and a male gorilla can weigh up to 500 pounds without being obese.
10. Nishihara, "Feeding Ecology of Western Lowland Gorillas in the Nouabale-Ndoki National Park, Congo."
11. Mori, "Comparison of the Communicative Vocalizations and Behaviors of Group Ranging in Eastern Gorillas, Chimpanzees and Pygmy Chimpanzees."
12. Fossey, "Vocalizations of the Mountain Gorilla (*Gorilla gorilla beringei*)."
13. Koko would have scored higher, but the test was shown to be culturally biased. For example, when Koko was asked where you go when it's raining, she was given a choice of a tree, a house, and another object that was totally unrelated. She chose "tree," which is absolutely correct if you are a gorilla. It was marked wrong on the test. For more on Koko's language capabilities, see Patterson, "The Gestures of a Gorilla: Language Acquisition in Another Pongid."
14. For a description of the stages of these displays and common contexts in which they occur, see Fossey, *Gorillas in the Mist*, pp. 54–57, 62–63, 66–69, 112–113, 184–185.
15. Ibid., p. 91.
16. Ibid., pp. 30, 31, 71, 220.
17. Ibid., pp. 47–48.
18. Ibid., p. 180.
19. Schaller quoted in Bourne, *Gentle Giants*, pp. 151–154.
20. Scientists disagree on this point. Some paleoanthropologists, for instance,

when finding a large fossil hominid specimen and a smaller one in the same site will classify them as two different species. Other scientists would be more likely to see the specimens as sexually dimorphous members of the same species.

21. This kind of "fashion sense" or interest in personal adornment has been seen in chimpanzees, also. Roger and Debbie Fouts personally communicated to me that the chimpanzees at Central Washington University's Research Center pick out favorite clothing items when they are introduced for enrichment and wear them.

22. Fossey, *Gorillas in the Mist*, p. 52. See also Goodall, *In the Shadow of Man*, pp. 171, 206.

23. Fossey, *Gorillas in the Mist*, p. 104.

24. See Yamagiwa, "Activity Rhythm and the Ranging of a Solitary Male Mountain Gorilla (*Gorilla gorilla beringei*)" and Fossey, "Reproduction among Free-Living Mountain Gorillas."

25. Fossey, "Imperiled Giants of the Forest."

26. Interestingly, very young gorillas have a small white flame of hair on their bottom that disappears at about four years of age. During the time that the white patch is evident, they are never disciplined roughly. Fossey, *Gorillas in the Mist*.

27. Fossey, "Development of the Mountain Gorilla (*Gorilla gorilla beringei*)," pp. 139–184.

28. The process of separate nesting for the young is long and involved and almost looks like cell division: the parent nest will move from an oval structure to having a bulge in one side to accommodate the growing offspring. Eventually, this bulge in the side of the nest will become separate, but the rims of the nests, like cell walls, will still touch. The final stage involves the maturing offspring's building a nest some distance from that of its mother.

29. Goodall, *In the Shadow of Man*, pp. 52–54.

30. Eisler, *Sacred Pleasure*.

31. Mori, "Comparison of the Communicative Vocalizations and Behaviors of Group Ranging in Eastern Gorillas, Chimpanzees, and Pygmy Chimpanzees."

32. In one particularly memorable instance, I saw a group of gorillas, all holding short sticks, sit in a perfect circle, all of them facing the center. They stayed in this arrangement for several minutes, absolutely silent. What the purpose of this gathering was, if there even was one, is impossible to say; but the impact was striking. See Campbell, *Primitive Mythology*, pp. 358–359.

33. Sally Boysen, personal communication.

34. Bourne, *Gentle Giants*, pp. 60–64.

35. Fossey, *Gorillas in the Mist*, p. 47.

36. Groves and Sabater Pi, "From Ape's Nest to Human Fix-Point."

37. Leakey, *Olduvai Gorge*.

38. Fossey, *Gorillas in the Mist*, p. 71.

39. Madam Bee's daughters had also been concerned with their mother's health and well-being before this: they routinely climbed tall trees to harvest fruit for her in her old age. Goodall, *Through a Window*, pp. 106–107, 211.

40. Goodall, *In the Shadow of Man*, pp. 218–224.

41. For touching examples of care of orphaned chimpanzees, see ibid., pp. 226–231.

42. Naturally occurring antibiotics can be found in such unlikely places as bacteria, fungi, algae, lichens, maggots, insects, blood, pig's feet, and snail slime. For an interesting paper on the subject, see Keith and Armelagos, "Naturally Occurring Dietary Antibiotics and Human Health."

43. Dorson, *Folk Legends of Japan*, p. 83.

44. Colarusso, "Wild Man of the Caucasus," pp. 262–263.

45. Tomitani et al., "Chlorophyll b and Phycobilins in the Common Ancestor of Cyanobacteria and Chloroplasts."

46. Bourne, *Gentle Giants*; Sabater Pi, "An Albino Lowland Gorilla from Rio Muni, West Africa, and Notes on Its Adaptation to Captivity."

47. Millar, *The Piltdown Men*.

48. Ross, *Pagan Celtic Britain*, pp. 60, 318, 321.

49. Hull, *Sun Dancing*.

50. Bradford and Blume, *Ota Benga*, pp. 38–58; Green, "Edwardian Britain's Forest Pygmies."

51. Newton, *Principia Mathematica*.

52. Trefil, "Architects of Time."

53. Klinkenborg, "The Best Clock in the World."

More about Gorillas

1. For an excellent approach to providing lay readers of all ages a thorough but clear account of primate evolution, see Zihlman, *The Human Evolution Coloring Book*.

2. For a more in-depth analysis of this specific period of primate evolution, see Rose and Fleagle, "The Fossil History of Non-Human Primates in the Americas" and Rosenberger and Szalay, "On the Tarsiiform Origins of Anthropoidea." The reader is also encouraged to read the recent paper by Gebo et al., "The Oldest Known Anthropoid Postcranial Fossils and the Evolution of Higher Primates."

3. For a concise overview of Platyrrhine origins and development and explorations of the Asian-origin and African-origin theories as well as a treatment of phylogenetic/cladistic and vicariance models, see Ciochon and Chiarelli, "Paleobiogeographic Perspectives on the Origin of the Platyrrhini."

4. See Gribbin and Cherfas, "Descent of Man—or Ascent of Ape?"

5. Wilson, Carlson, and White, "Biochemical Evolution."

6. The average weight for *G. g. gorilla* males is 140 kg, and for females, 75 kg; for *G. g. graueri* males, 165 kg, and for females, 80 kg; for *G. g. beringei* males, 160 kg, and for females, 85 kg. See Estes, *The Behavior Guide to African Mammals*, p. 535.

7. The hair of *G. g. beringei* is long, black, and very dense, suitable to the colder climate in which it lives; the hair of both *G. g. gorilla* and *G. g. graueri* is less dense, with a somewhat wider range of color. See ibid.

8. *G. g. graueri* have the thinnest face; *G. g. gorilla* have a face of medium width; *G. g. beringei* have a massively wide face. See ibid.

9. Casimir, "Feeding Ecology of an Eastern Gorilla Group in the Mt. Kahuzi Region (Republique du Zaire)."

10. Of course, this practice also opens up roads into the gorillas' ranges, which may not be good for them in the long run.

11. Tutin, "Ranging and Social Structure of Lowland Gorillas in the Lope Reserve, Gabon."

12. Watts, "Long-Term Habitat Use by Mountain Gorillas (*Gorilla gorilla beringei*)."

13. Estes, *The Behavior Guide to African Mammals*, p. 535.

14. Ibid.

15. For a straightforward description of these universal cultural components, see Whiteford, *The Human Portrait*, p. 63.

16. Gorilla population estimates are from 1994 American Zoo and Aquarium Association statistics.

References and Further Reading

Andrews, Peter. "Evolution and Environment in the Hominoidea." *Nature* 360 (1992): 641–646.

Argyle, Michael. *Bodily Communication*. New York: Methuen & Co., 1994.

Armstrong, D. M. *A Theory of Universals*, vol. 2. Cambridge, U.K.: Cambridge University Press, 1978.

Badrian, Alison, and Noel Badrian. "Social Organization of *Pan paniscus* in the Lomako Forest, Zaire." Pp. 325–344 in *The Pygmy Chimpanzee: Evolutionary Biology and Behavior*, ed. Randall L. Susman. New York: Plenum Press, 1984.

Bard, Kim. "Evolutionary Foundations of Intuitive Parenting: Maternal Competence in Chimpanzees." *Early Development and Parenting* 3 (1994): 19–28.

Beck, Benjamin. *Animal Tool Behavior: The Use and Manufacture of Tools by Animals*. New York: Garland STPM Press, 1980.

Bernor, R. L. "Geochronology and Zoogeographic Relationships of Miocene Hominoidea." Pp. 21–26 in *New Interpretations of Ape and Human Ancestry*, ed. Russell L. Ciochon and Robert S. Coruccini. New York: Plenum Press, 1983.

Bradford, Phillips Vernor, and Harvey Blume. *Ota Benga: The Pygmy in the Zoo*. New York: St. Martins Press, 1992.

Brown, Donald E. *Human Universals*. Philadelphia: Temple University Press, 1991.

Brown, W. M., E. M. Prager, A. Wang, and A. C. Wilson. "Mitochondrial DNA Sequences of Primates: Tempo and Mode of Evolution." *Journal of Molecular Evolution* 18 (1982): 225–239.

Bourne, Geoffrey. *The Gentle Giants: The Gorilla Story*. New York: G. P. Putnam's Sons, 1975.

Byrne, Richard. "The Smart Gorilla's Recipe Book." *Natural History* 10 (1995): 13–15.

Campbell, Joseph. *Primitive Mythology: The Masks of God*. New York: Viking Penguin, 1959.

Casimir, M. J. "Feeding Ecology of an Eastern Gorilla Group in the Mt. Kahuzi Region (Republique du Zaire)." *Folia Primatologica* 24 (1975): 81–136.

Cheney, Dorothy L., and Robert Seyfarth. *How Monkeys See the World*. Chicago: Chicago University Press, 1990.

Chiarelli, A. B. *Evolution of the Primates*. New York: Academic Press, 1973.

Ciochon, R. L., and A. B. Chiarelli. "Paleobiogeographic Perspectives on the Origin of the Platyrrhini." Pp. 110–123 in *Primate Evolution*, ed. R. L. Ciochon and J. G. Fleagle. Menlo Park, Calif.: The Benjamin/Cummings Publishing Company, 1985.

Clark, W. E. LeGros, *History of the Primates*. Chicago: University of Chicago Press, 1965.

Colarusso, John. "A Wild Man of the Caucasus." Pp. 262–263 in *Manlike Monsters on Trial*, ed. Marjorie Halpin. Vancouver: University of British Columbia Press, 1980.

Conroy, Glenn C., and David Pilbeam. "Ramapithecus: A Review of Its Hominid

Status." Pp. 59–86 in *Paleoanthropology: Morphology and Paleoecology*, ed. Russell E. Tuttle. The Hague: Mouton Publishers, 1975.

Crelin, Edmund S. *The Human Vocal Tract: Anatomy, Function, Development, and Evolution.* New York: Vantage Press, 1987.

Dart, Raymond. "The Osteodontokeratic Culture of *Australopithecus africanus*." *Memoirs of the Transvaal Museum* 10 (1957): 1–105.

DeVore, Irven, and Sarel Eimerl. *The Primates.* New York: Time-Life Books, 1965.

Dorson, Richard M. *Folk Legends of Japan.* New York: Charles E. Tuttle Co., 1962.

Ehrenberg, Margaret. *Women in Prehistory.* Norman: University of Oklahoma Press, 1989.

Eisler, R. Jane. *Sacred Pleasure: Sex, Myth, and the Politics of the Body.* San Francisco: Harper San Francisco, 1995.

Estes, Richard Despard. *The Behavior Guide to African Mammals.* Berkeley: University of California Press, 1991.

Fossey, Dian. "Vocalizations of the Mountain Gorilla (*Gorilla gorilla beringei*)." *Animal Behavior* 20 (1972): 36–53.

———. "Development of the Mountain Gorilla (*Gorilla gorilla beringei*): The First 36 Months." Pp. 130–185 in *The Great Apes: Perspectives on Human Evolution*, vol. 5, ed. David A. Hamburg and Elizabeth R. McCowan. Menlo Park, Calif.: The Benjamin/Cummings Publishing Company, 1979.

———. "Imperiled Giants of the Forest." *National Geographic* 4 (1981): 501–523.

———. "Reproduction among Free-Living Mountain Gorillas." *American Journal of Primatology* Supplement 1 (1982): 97–104.

———. *Gorillas in the Mist.* Boston: Houghton Mifflin Company, 1983.

Freudenberg, Wilhelm. "Die Entdeckung von Menlischen Fubspuren und Artefact in den Tertiaren Gerolschichten und Muschelhaufen bei St. Gilles-Waes, Westlich Antwerpen," trans. Michael Cremo and Richard L. Thompson. *Preahistorische Zeitschrift* 11 (1919).

Frisby, John. *Seeing.* Oxford: Oxford University Press, 1979.

Galdikas, Birute. *Reflections of Eden: My Years with the Orangutans of Borneo.* New York: Little, Brown and Company, 1995.

Gebo, Daniel L., Marian Dagosto, K. Christopher Beard, Tao Qi, and Jingwen Wang. "The Oldest Known Anthropoid Postcranial Fossils and the Evolution of Higher Primates." *Nature* 3 (2000): 276–278.

Goodall, Jane. "Chimpanzees of the Gombe Stream Reserve." Pp. 425–473 in *Primate Behavior: Field Studies of Monkeys and Apes*, ed. Irven DeVore. New York: Holt, Rinehart and Winston, 1965.

———. *My Friends the Wild Chimpanzees.* Washington, D.C.: National Geographic Society, 1967.

———. *In the Shadow of Man.* Boston: Houghton Mifflin Company, 1971.

———. *Through a Window: My Thirty Years with the Chimpanzees of Gombe.* Boston: Houghton Mifflin Company, 1990.

Green, Jeffrey. "Edwardian Britain's Forest Pygmies." *History Today* 45, no. 8 (August 1995).

Green, John. *Sasquatch: The Apes among Us.* Saanitchton, B.C.: Hancock House Publishers, 1978.

Gribbin, John, and Jeremy Cherfas. "Descent of Man—or Ascent of Ape?" *New Scientist* 3 (1981): 592–595.

Groves, Colin P., and J. Sabater Pi. "From Ape's Nest to Human Fix-Point." *Man* 20 (1985): 22–47.

Harcourt, A. H., and S. A. Harcourt. "Insectivory by Gorillas." *Folia Primatologica* 43 (1984): 229–233.

Hofstadter, Douglas F. "Variations on a Theme as the Crux of Creativity." Pp. 232–259 in Hofstadter, *Metamagical Themas: Questing for the Essence of Mind and Matter*. New York: Bantam Books, 1985.

Howell, F. Clark. *Early Man*. New York: Time-Life Books, 1965.

Hrdy, Sarah Blaffer. *The Woman That Never Evolved*. Harvard: Harvard University Press, 1981.

Huffman, Michael A. "Tool-Assisted Predation on a Squirrel by a Female Chimpanzee in the Mahale Mountains, Tanzania." *Primates* 34 (1993): 93–98.

Hulbert, Katherine W. "Hominoid-Hominid Heterography and Evolutionary Patterns." Pp. 153–161 in *Paleoanthropology: Morphology and Paleoecology*, ed. Russell H. Tuttle. The Hague: Mouton Publishers, 1975.

Hull, Michael. *Sun Dancing: A Spiritual Journey on the Red Road*. Rochester, Vt.: Inner Traditions International, 2000.

Itani, Junichiro. "Social Structures of African Great Apes." *Journal of Reproduction and Fertility* 28 (1980): 33–41.

Janson, H. W. *Apes and Ape Lore in the Middle Ages and the Renaissance*. London: University of London Press, 1952.

Kano, Takayoshi. "The Use of Leafy Twigs for Rain Cover by the Pygmy Chimpanzees of Wamba." *Primates* 23 (1982): 253–257.

———. "An Ecological Study of the Pygmy Chimpanzees (*Pan paniscus*) of Yalosidi, Republic of Zaire." *International Journal of Primatology* 4 (1983): 1–31.

———. *The Last Ape: Pygmy Chimpanzee Behavior and Ecology*. Stanford: Stanford University Press, 1992.

Kay, R. F., and E. L. Simons. "A Reassessment of the Relationship between Later Miocene and Subsequent Hominoidea." Pp. 181–238 in *New Interpretations of Ape and Human Ancestry*, ed. Russell L. Ciochon and Robert S. Coruccini. New York: Plenum Press, 1983.

Keddie, Grant R. "On Creating Unhumans." Pp. 22–30 in *The Saquatch and Other Unknown Hominoids*, ed. Vladimir Markotic. Calgary: Western Publishers, 1984.

Keith, Margaret, and George Armelagos. "Naturally Occurring Dietary Antibiotics and Human Health." Pp. 221–230 in *The Anthropology of Medicine: From Culture to Method*, ed. Lola Romanucci-Ross, Daniel E. Moerman, and Laurence R. Tancredi. New York: J. F. Bergin Publishers, 1983.

King, Barbara J. *The Information Continuum*. Santa Fe: School of American Research Press, 1994.

Klinkenborg, Verlyn. "The Best Clock in the World." *Discover* 21, no. 6 (June 2000): 50.

Koffman, Jeanne. "A Brief Ecological Description of the Caucasus Relic Hominoid (Almasti) Based on Oral Reports by Local Inhabitants and on Field Investigations." Pp. 76–86 in *The Sasquatch and Other Unknown Hominoids*, ed. Vladimir Markotic. Calgary: Western Publishers, 1984.

Kortlandt, A. "Facts and Fallacies Concerning Miocene Ape Habitats." Pp. 465–514 in *New Interpretations of Ape and Human Ancestry*, ed. Russell L. Ciochon and Robert S. Coruccini. New York: Plenum Press, 1983.

Kuroda, Suehisa, Shigeru Suzuki, and Tomoaki Nishhara. "Preliminary Report on Predatory Behavior and Meat Sharing in Tschego Chimpanzees (*Pan troglodytes troglodytes*) in the Ndoki Forest, Northern Congo." *Primates* 37 (1996): 253–259.

Leakey, M. D. *Olduvai Gorge*, vol. 3. Cambridge, U.K.: Cambridge University Press, 1971.

Leakey, M. D., and J. M. Harris, eds. *Laetoli: A Pliocene Site in Northern Tanzania.* New York: Oxford University Press, 1977.

Linden, Eugene. *Apes, Men, and Language.* Chicago: University of Chicago Press, 1962.

———. "Chimpanzees with a Difference: Bonobos." *National Geographic* 3 (1992): 46–53.

Martin, R. D. *Primate Origins and Evolution: A Phylogenetic Reconstruction.* New Jersey: Princeton University Press, 1990.

Miller, Ronald. *The Piltdown Men.* New York: Ballantine, 1974.

Mori, Akio. "Comparison of the Communicative Vocalizations and Behaviors of Group Ranging in Eastern Gorillas, Chimpanzees, and Pygmy Chimpanzees." *Primates* 24 (1983): 486–500.

Moynihan, Martin. *The New World Primates: Adaptive Radiation and the Evolution of Social Behavior, Languages, and Intelligence.* New Jersey: Princeton University Press, 1976.

Nelson, Harry, Robert Jurmain, and Lynn Kilgore. *Essentials of Physical Anthropology.* St. Paul: West Publishing Company, 1992.

Newton, Isaac. *Philosophiae Naturalis Principia Mathematica.* Cambridge, Mass.: Harvard University Press, 1972.

Nishihara, Tomoaki. "A Preliminary Report on the Feeding Habits of Western Lowland Gorillas (*Gorilla gorilla gorilla*) in the Ndoki Forest, Northern Congo." Pp. 225–233 in *Topics in Primatology*, ed. Itoigawa et al. Tokyo: University of Tokyo Press, 1991.

———. "Feeding Ecology of Western Lowland Gorillas in the Nouabale-Ndoki National Park, Congo." *Primates* 36 (1995): 151–168.

Patnaik, Rejeev, and David Cameron. "New Miocene Fossil Ape Locality, Dangar, Hari-Talyangar Region, Siwaliks, Northern India." *Journal of Human Evolution* 32 (1997): 93–97.

Patterson, F. "The Gestures of a Gorilla: Language Acquisition in Another Pongid." *Brain and Language* 5 (1978): 72–97.

Peterson, Dale, and Jane Goodall. *Visions of Caliban: On Chimpanzees and People.* New York: Houghton Mifflin Company, 1993.

Pfeiffer, John. *The Emergence of Man.* New York: Harper & Row Publishers, 1969.

Pilbeam, D. R. "The Earliest Hominoids." Pp. 211–214 in *Primate Evolution and Human Origins*, ed. Russell L. Ciochon and John G. Fleagle. New York: Aldine De Gruyer, 1987.

Plavcan, Michael J., and Carel P. Van Schaik. "Interpreting Hominid Behavior on the Basis of Sexual Dimorphism." *Journal of Human Evolution* 32 (1997): 345–374.

Prasad, K. N. "Was Ramapithecus a Tool User?" *Journal of Human Evolution* 11 (1982): 101–104.

Prince-Hughes, Dawn. *The Archetype of the Ape-Man: The Phenomenological Excavation of a Relic Hominid Ancestor.* Ph.D. dissertation, Universitat Herisau. Available at dissertation.com.

Radetsky, Peter. "Silence, Signs, and Wonder." *Discover* 8 (1994): 60–68.

Renfrew, Colin, and Paul Bahn. *Archaeology: Theories, Methods and Practice.* New York: Thames and Hudson, 1991.

Retallack, Gregory S., D. P. Dugas, and E. A. Bestland. "Fossil Soils and Grasses of a Middle Miocene East African Grassland." *Science* 247 (1990): 1325–1328.

Reynolds, Vernon, and Frances Reynolds. "Chimpanzees of the Budongo Forest." Pp. 368–424 in *Primate Behavior: Field Studies of Monkeys and Apes,* ed. Irven DeVore. New York: Holt, Rinehart and Winston, 1965.

Robbins, L. M. "Hominid Footprints from Site G." Pp. 497–502 in *Laetoli: A Pliocene Site in Northern Tanzania,* ed. M. D. Leakey and J. M. Harris. New York: Oxford University Press, 1977.

Rodgers, Elizabeth M., Fiona Maisels, Elizabeth A. Williamson, Michael Fernandez, and Caroline E. G. Tutin. "Gorilla Diet in the Lope Reserve, Gabon: A Nutritional Analysis." *Ecologia* 84 (1990): 326–339.

Rose, K. D., and J. G. Fleagle. "The Fossil History of Non-Human Primates in the Americas." Pp. 111–167 in *Ecology and Behavior of Neotropical Primates,* vol. 1, ed. A. F. Coimbra-Filho and R. A. Mittermeier. Rio de Janiero: Academia Brasiliera de Ciencias, 1981.

Rosenberger, A. L., and F. S. Szalay. "On the Tarsiiform Origins of Anthropoidea." Pp. 139–157 in *Evolutionary Biology of the New World Monkeys and Continental Drift,* ed. R. L. Ciochon and A. B. Chiarelli. New York: Plenum Press, 1980.

Ross, Anne. *Pagan Celtic Britain.* Chicago: Academy Chicago Publishers, 1996.

Sabater Pi, J. "An Albino Lowland Gorilla from Rio Muni, West Africa, and Notes on Its Adaptation to Captivity." *Folia Primatologica* 7 (1967): 155–160.

Savage, R. J. G. *Mammal Evolution.* New York: Facts on File Publications, 1990.

Schaller, George B. *The Year of the Gorilla.* New York: Ballantine Books, 1964.

———. "Behavior of the Mountain Gorilla." Pp. 324–367 in *Primate Behavior: Field Studies of Monkeys and Apes,* ed. Irven DeVore. New York: Holt, Rinehart and Winston, 1965.

Schick, Kathy D., and Nikolas Toth. *Making Silent Stones Speak: Human Evolution and the Dawn of Technology.* New York: Simon and Schuster, 1993.

Sheets-Johnstone, Maxine. *The Roots of Thinking.* Philadelphia: Temple University Press, 1990.

Simons, E. L., and D. R. Pilbeam. "Hominoid Paleoprimatology." Pp. 36–62 in *The Functional and Evolutionary Biology of Primates,* ed. R. Tuttle. Chicago: Aldine, 1972.

Singer, Peter. "The Significance of Animal Suffering." Pp. 233–244 in *Ethical Issues in Scientific Research: An Anthology,* ed. Edwin Erwin, Sidney Gendin, and Lowell Klienman. New York: Garland Publishing, 1994.

Spuhler, J. N. "Evolution of Mitochondrial DNA in Monkeys, Apes, and Humans." *Yearbook of Physical Anthropology* 31 (1988): 15–48.

Susman, Randall L. "The Locomotor Behavior of *Pan paniscus* in the Lomako Forest." Pp. 369–394 in *The Pygmy Chimpanzee: Evolution, Biology, and Behavior*, ed. Randall L. Susman. New York: Plenum Press, 1984.

Tomitani, Akiko, Kiyotaka Okada, Hideaki Miyashita, Hans C. P. Matthijs, Terufumi Ohno, and Ayumi Tanaka. "Chlorophyll b and Phycobilins in the Common Ancestor of Cyanobacteria and Chloroplasts." *Nature* 400, no. 6740 (1999): 159.

Trefil, James. *Sharks Have No Bones: 1001 Things Everyone Should Know about Science*. New York: Fireside, 1992.

———. "Architects of Time." *Astronomy* 27, no. 9 (September 1999).

Tutin, Caroline. "Ranging and Social Structure of Lowland Gorillas in the Lope Reserve, Gabon." Pp. 58–70 in *Great Ape Societies*, ed. W. C. McGrew, L. F. Marchant, and T. Nishida. Cambridge, U.K.: Cambridge University Press, 1996.

Tutin, Caroline, and Michael Fernandez. "Gorillas Feeding on Termites in Gabon, West Africa." *Journal of Mammalogy* 64 (1983): 530–531.

———. "Insect Eating by Sympatric Lowland Gorillas (*Gorilla gorilla gorilla*) and Chimpanzees (*Pan t. troglodytes*) in the Lope Reserve, Gabon." *American Journal of Primatology* 28 (1992): 29–40.

Waal, F. de. "Observations on the Ontogeny of Feeding Behavior in Mountain Gorillas (*Gorilla gorilla beringei*)." *American Journal of Primatology* 8 (1985): 1–10.

Watts, David P. "Long-Term Habitat Use by Mountain Gorillas (*Gorilla gorilla beringei*). 1. Consistency, Variation, and Home Range Size and Stability." *International Journal of Primatology* 19 (1994): 651–680.

Whiteford, Michael. *The Human Portrait: Introduction to Cultural Anthropology*. Englewood Cliffs: Prentice Hall, 1992.

Willis, Delta. *The Hominid Gang: Behind the Scenes in the Search for Human Origins*. New York: Viking, 1989.

Wilson, A. C., S. S. Carlson, and T. J. White. "Biochemical Evolution." *Annual Review of Biochemistry* 46 (1977): 573–639.

Yamagiwa, J. "Activity Rhythm and the Ranging of a Solitary Male Mountain Gorilla (*Gorilla gorilla beringei*)." *Primates* 27 (1986): 273–282.

Zihlman, Adrienne. *The Human Evolution Coloring Book*. New York: HarperCollins Publishers, 1982.

About the Author

Dawn Prince-Hughes grew up in the wilderness of northwestern Montana, where she learned to appreciate animals and their habitats and the ways in which they nourish the archaic parts of the human soul. She combined zoo medicine and primate behavior studies at Woodland Park Zoo in Seattle with her educational program and completed a Ph.D. in interdisciplinary anthropology in 1997 at Universitat Herisau in Switzerland. She works on behalf of primates in both captive and natural settings as a Research Associate at Western Washington University and through her support of the Jane Goodall Institute's Chimpanzoo program. Dr. Prince-Hughes currently lives with her partner and their son in Bellingham, Washington, where she writes both academic works and novels on apes, emergent cultures, human origins, and the importance of our ancient connections.